Titu Andreescu
Zuming Feng

A Path to Combinatorics
for Undergraduates
Counting Strategies

Birkhäuser
Boston • Basel • Berlin

Titu Andreescu
American Mathematics Competitions
University of Nebraska
Lincoln, NE 68588
U.S.A.

Zuming Feng
Phillips Exeter Academy
Department of Mathematics
Exeter, NH 03833
U.S.A.

Library of Congress Cataloging-in-Publication Data

Andreescu, Titu, 1956-
   A path to combinatorics for undergraduates : counting strategies / Titu Andreescu,
Zuming Feng.
      p. cm.
   Includes bibliographical references and index.
   ISBN 0-8176-4288-9 (alk. paper) — ISBN 3-7643-4288-9 (alk. paper)
      1. Combinatorial analysis. 2. Combinatorial number theory. I. Feng, Zuming. II. Title.

QA164.A58 2003
511'.6–dc22   2003057761
                                                                              CIP

AMS Subject Classifications: 0501, 05A15, 05A10, 05A05, 05A19, 60C05, 03D20

Printed on acid-free paper.
©2004 Birkhäuser Boston   *Birkhäuser*

ISBN 0-8176-4288-9   SPIN 10864464
ISBN 3-7643-4288-9

Cover design by Joseph Sherman.
Printed in the United States of America.

9 8 7 6 5 4 3 2 1

Birkhäuser   Boston • Basel • Berlin
A member of BertelsmannSpringer Science+Business Media GmbH

Contents

Preface

The main goal of the two authors is to help undergraduate students understand the concepts and ideas of combinatorics, an important realm of mathematics, and to enable them to ultimately achieve excellence in this field. This goal is accomplished by familiarizing students with typical examples illustrating central mathematical facts, and by challenging students with a number of carefully selected problems. It is essential that the student works through the exercises in order to build a bridge between ordinary high school permutation and combination exercises and more sophisticated, intricate, and abstract concepts and problems in undergraduate combinatorics. The extensive discussions of the solutions are a key part of the learning process.

The concepts are not stacked at the beginning of each section in a blue box, as in many undergraduate textbooks. Instead, the key mathematical ideas are carefully worked into organized, challenging, and instructive examples. The authors are proud of their strength, their collection of beautiful problems, which they have accumulated through years of work preparing students for the International Mathematics Olympiads and other competitions.

A good foundation in combinatorics is provided in the first six chapters of this book. While most of the problems in the first six chapters are real counting problems, it is in chapters seven and eight where readers are introduced to essay-type proofs. This is the place to develop significant problem-solving experience, and to learn when and how to use available skills to complete the proofs.

Introduction

Combinatorics is a classical branch of mathematics. Although combinatorics represents a special case among other branches of mathematics, mostly because there is no axiomatic theory for it, some of the greatest mathematicians studied combinatorial problems. By axiomatic classification we might think of combinatorics to be a chapter of the theory of numbers, but clearly, combinatorics deals with totally different kinds of problems. Combinatorics interacts with other mathematical theories, from which it takes problem-solving strategies and to which it provides results. Combinatorics deals with concrete problems, easy to understand and often presented in a form suitable for publication in a regular newspaper, as opposed to high-end abstract mathematics journals. This characteristic makes it attractive to both beginners and professional mathematicians. These kinds of problems are particularly useful when one learns or teaches mathematics. Here is one example:

Example 0.1. *A town is rectangularly shaped, and its street network consists of $x + 1$ parallel lines headed north–south and $y + 1$ parallel lines headed east–west. (You may think of Manhattan as an example.) In how many ways can a car reach the northeast corner if it starts in the southwest corner and travels only in the east and north directions?*

Solution: Let A and B represent the southwest and northeast corners, respectively. In the following figure we show such a grid for $x = 5$ and $y = 4$.

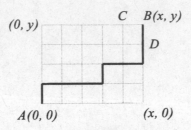

Figure 0.1.

We denote by $v_0, v_1, \ldots, v_x$ the vertical (north-south) streets. Let us call a *block* the interval between two neighboring horizontal streets. (Note that under this definition, a block can be only a part of a vertical street.) With a bit of thought, we see that a path that connects the points A and B is completely determined by the number of blocks along the path in each of the vertical streets $v_0, v_1, \ldots, v_x$. For $0 \leq i \leq x$, let y_i be the number of such blocks for street v_i. Then y_i are nonnegative integers and

$$y_0 + y_1 + y_2 + \cdots + y_x = y. \qquad (*)$$

Hence, the number of paths connecting the points A and B is equal to the number of ordered $(x+1)$-tuples $(y_0, y_1, \ldots, y_x)$ of nonnegative integers satisfying the equation $(*)$. (Indeed, Figure 0.1 corresponds to the solution $(y_0, y_1, y_2, y_3, y_4, y_5) = (1, 0, 0, 1, 0, 2)$ for the equation $y_0 + y_1 + \cdots + y_5 = 4$.) Thus, we rediscover the problem of de Moivre (a classic problem in arithmetic): In how many ways can a nonnegative integer y be written as a sum of $x + 1$ nonnegative integers?

For an algebraist, this problem is equivalent to the following: What is the number of unitary monomials $Z_0^{y_0} Z_1^{y_1} \cdots Z_x^{y_x}$ in $x+1$ variables, of degree d, contained in the ring of polynomials $\mathbb{R}[Z_0, Z_1, \ldots, Z_x]$? It is useful to know this number, since it is equal to the dimension of the vector space of homogeneous polynomials (homogeneous forms) of degree y.

We now have two different characterizations of the desired number, but we still do not know its value. Let us try something else!

Let $N_{x,y}$ be the number we want to find. One can see that $N_{1,y} = y + 1$. Putting the grid into a coordinate plane with $A = (0,0)$ and $B = (x, y)$, we see that $N_{0,1} = N_{1,0} = 1$, $N_{0,2} = 1$, $N_{1,1} = 2$, $N_{2,0} = 1$, and so on. More generally, since every path from A to $B = (x, y)$ passes through $C = (x - 1, y)$ or $D = (x, y - 1)$, one can obtain the recursive relation

$$N_{x,y} = N_{x,y-1} + N_{x-1,y}. \qquad (**)$$

Again, for an algebraist, the same relation can be obtained by noticing that every monomial of degree y in $Z_1, \ldots, Z_x$ either is a monomial of degree y in $Z_0, Z_1, \ldots, Z_{x-1}$ or is obtained by multiplying a monomial of degree $y - 1$ in $Z_0, Z_1, \ldots, Z_{x-1}$ by Z_x.

This *recursive* approach is promising. It clearly resembles the famous **Pascal's triangle**. By induction, we obtain

$$N_{x,y} = \binom{x+y}{x} = \binom{x+y}{y} = \frac{(x+y)!}{x! \cdot y!}.$$

(It also common to write $N_{x,y} = {}_{(x+y)}C_x = {}_{(x+y)}C_y = C(x+y, x) = C(x+y, y)$.) We thus obtain the following result: The number of paths connecting the points A and B is equal to the number of ordered sequences of $x + 1$ nonnegative integers that add up to y, and is also equal to the number of unitary monomials of degree y in $x + 1$ variables, and, finally, is equal to $\binom{x+y}{x}$. ∎

Thinking back to the way we obtained this result, we observe the following principle:

> *In order to count the elements of a certain set, we might replace the original set with another set that has the same number of elements and whose elements can be more easily counted.*

With the above principle in mind, we present two new ways of solving the above problem.

First, we notice that it takes exactly $x + y$ moves to get from A to B. Among those moves, x of them have to be going right, and y of them have to be going up. There is a bijective mapping between the set of paths connecting the points A and B and the set of distinct arrangements of x R's and y U's. For example, the sequence $URRRURRUU$ corresponds to the path in Figure 0.1. Hence, $N_{x,y} = \binom{x+y}{x} = \binom{x+y}{y}$.

Second, we can consider $x + y$ identical white balls lined up in a row. We pick x balls and color them red. Label the red balls $r_1, r_2, \ldots, r_x$ from left to right in that order. For $0 \leq i \leq x$, let y_i denote the number of white balls in between red balls r_i and r_{i+1}. (Here y_0 and y_x denote the numbers of balls to the left of ball r_1 and to the right of ball r_x, respectively.) There is a bijective mapping between the set of all such colorings and the set of $(x+1)$-tuples of nonnegative integer solutions to equation $(*)$. For example, for $x = 5$ and $y = 4$, the coloring in Figure 0.2 corresponds to $(y_0, y_1, y_2, y_3, y_4, y_5) = (1, 0, 0, 1, 0, 2)$ for $y_0 + y_1 + \cdots + y_5 = 4$.

Figure 0.2.

Indeed, the red balls become separators for the remaining y white balls. The number of consecutive white balls between two red balls maps to a unique solution of $(*)$ and vice versa. It is clear that there are $\binom{x+y}{x} = \binom{x+y}{y}$ such colorings.   ■

Example 0.2.   [Balkan 1997] *Let m and n be integers greater than 1. Let S be a set with n elements, and let $A_1, A_2, \ldots, A_m$ be subsets of S. Assume that for any two elements x and y in S, there is a set A_i such that either x is in A_i and y is not in A_i or x is not in A_i and y is in A_i. Prove that $n \leq 2^m$.*

Solution: Let us associate with each element x in S a sequence of m binary digits $a(x) = (x_1, x_2, \ldots, x_m)$ such that $x_i = 1$ if x is in A_i and $x_i = 0$ if x is not in A_i. We obtain a function

$$f: \ S \to T = \{(x_1, \ldots, x_m) \mid x_i \in \{0, 1\}\},$$

and the hypothesis now states that if $x \neq y$, then $f(x) \neq f(y)$. (Such a function is called an injective function.) Thus, the set T must have at least as many elements as the set S has. It is not difficult to see that T has 2^m elements, because each component x_i of $(x_1, \ldots, x_m)$ can take exactly two possible values, namely, 0 and 1. It follows that $n \leq 2^m$.   ■

This problem also illustrates a solving strategy: In order to obtain a requested inequality, we might be able to transform the hypothesis about the family of subsets $(A_i)_{1 \leq i \leq m}$ by comparing the set S and the set of binary sequences of length m.

Example 0.3.   *An even number of persons are seated around a*

*table. After a break, they are again seated around the same table, not
necessarily in the same places. Prove that at least two persons have
the same number of persons between them as before the break.*

Solution: The problem is equivalent to the following: Given a
permutation $\sigma \in S_{2n}$, there exists a pair i, j such that

$$|i - j| = |\sigma(i) - \sigma(j)|.$$

Let us imagine the following construction: Suppose the $2n$ numbers
are the vertices of a regular polygon, $P_0, \ldots, P_{2n-1}$. The permutation
σ sends vertex P_i to $P_{\sigma(i)}$. The number of vertices between P_a and P_b
is the same after the permutation σ if $a - \sigma(a) \equiv b - \sigma(b) \,(\text{mod } 2n)$.
Suppose, by way of contradiction, that this is not true for any a, b.
Then the numbers $i - \sigma(i)$ must be $0, 1, 2, \ldots, 2n - 1 \,(\text{mod } 2n)$. Their
sum is $(2n - 1)n \,(\text{mod } 2n)$. On the other hand, the $\sigma(i)$'s are a
permutation of the i's, so the sum of the i's equals the sum of the
$\sigma(i)$'s. Thus, the sum of the numbers $i - \sigma(i)$ must be 0. This
contradicts the value of $(2n - 1)n$ deduced above.   ∎

> *In this problem, the main idea is to construct a model such
> that accessible mathematical tools might be used.*

In analyzing the above examples, the following question arises: Are
there general methods for counting the elements of a certain set? The
answer is generally yes, and combinatorics provides such methods and
also teaches us how to find them in order to be able to solve problems
like the ones above.

Acknowledgments

Many thanks to Emily Bernier, Charles Chen, Patricia Hindman, and Tony Zhang, who helped with proofreading and improved the writing of the text. Special thanks to Po-Ling Loh, Ioan Tomescu, and Zoran Šuniḱ for the careful proofreading of the final version of the manuscript.

Many problems were either inspired by or adapted from mathematical contests in different countries and from the following journals:

High-School Mathematics, China
Revista Matematică Timişoara, Romania

We did our best to cite all original sources of the problems in the solution section. We express our deepest appreciation to the proposers of the original problems.

Abbreviations and Notation

Abbreviations

AHSME	American High School Mathematics Examination
AIME	American Invitational Mathematics Examination
AMC10	American Mathematics Contest 10
AMC12	American Mathematics Contest 12, which replaced AHSME
ARML	American Regions Mathematics League
Balkan	Balkan Mathematics Olympiad
IMC	International Mathematics Competition (for university students)
IMO	International Mathematical Olympiad
A.M.M.	American Mathematical Monthly
MOSP	Mathematical Olympiad Summer Program
PEA	Phillips Exeter Academy
Putnam	William Lowell Putnam Mathematical Competition
USAMO	United States of America Mathematical Olympiad
TST	Team Selection Test (for the USA IMO Team)
St. Petersburg	St. Petersburg (Leningrad) Mathematical Olympiad

Notation for Numerical Sets and Fields

$\mathbb{Z}$	the set of integers
$\mathbb{Z}_n$	the set of integers modulo n
$\mathbb{N}$	the set of positive integers
$\mathbb{N}^0$	the set of nonnegative integers
$\mathbb{Q}$	the set of rational numbers
$\mathbb{Q}^+$	the set of positive rational numbers
$\mathbb{Q}^0$	the set of nonnegative rational numbers
$\mathbb{Q}^n$	the set of n-tuples of rational numbers
$\mathbb{R}$	the set of real numbers
$\mathbb{R}^+$	the set of positive real numbers
$\mathbb{R}^0$	the set of nonnegative real numbers
$\mathbb{R}^n$	the set of n-tuples of real numbers
$\mathbb{C}$	the set of complex numbers
$[x^n](p(x))$	the coefficient of the term x^n in the polynomial $p(x)$

Notation for Sets, Logic, and Geometry

$|A|$	the number of elements in the set A
$\overline{A}$	the complement of A
$A \subset B$	A is a proper subset of B
$A \subseteq B$	A is a subset of B
$A \setminus B$	A without B (set difference)
$A \cap B$	the intersection of sets A and B
$A \cup B$	the union of sets A and B
$a \in A$	the element a belongs to the set A

A Path to Combinatorics
for Undergraduates

1
Addition or Multiplication?

Counting is one of the most fundamental skills. People start to count on their fingers when they are in kindergarten or even earlier. But how to count quickly, correctly, and systematically is a lifelong course.

Example 1.1. Determine the maximum among all numbers obtained by deleting 100 digits from the number

$$12345678910111213\ldots99100,$$

whose digits are the integers 1 through 100 in order from left to right.

Solution: We start with some counting exercises. There are nine single digit numbers. From 10 to 99, there are $99 - 10 + 1 = 90$ two-digit numbers. Hence the given number has $9 + 2 \cdot 90 + 3 = 192$ digits. We will obtain a 92-digit number after deleting 100 digits. For any two numbers with the same number of digits, the number with higher leading digits is larger. Hence, our number should start with as many 9's as possible. For this, we first delete the 8 leftmost digits. We then delete the string $101112\ldots181$ for a total of $9 \times 2 + 1 = 19$ digits. Similarly, we delete strings $202122\ldots282$, $303132\ldots383$, $404142\ldots484$. Hence we have deleted $8 + 19 \times 4 = 84$ digits, and we end up with the number

$$9999950515253\ldots99100. \qquad (*)$$

We still need to delete 16 digits. This seems to be a no-brainier: Just delete the string $505152\ldots57$ for a total 16 digits to obtain the

1

number

$$9999958596061\ldots99100.$$

Not so fast, my friend! (This is our favorite quotation from ESPN college football commentator Lee Croso.) If we keep the five leading 9's, the largest possible value for the next digit is 7, obtained by deleting the string $505152\ldots565$ for a total of 15 digits. The last digit to delete is the 5 from 58. Hence the answer is

$$9999978596061\ldots99100.$$

■

Example 1.2. Professors Alpha, Beta, Gamma, and Delta are giving graduate student Pi an oral qualifying exam on combinatorics. The professors are sitting in a row of four chairs. As the co-chairs of the examination committee, professors Alpha and Delta need to sit next to each other. As the adviser of student Pi, professor Gamma is required to sit next to the exam co-chairs. In how many ways can the professors sit?

Solution: The number of seating choices for professor Gamma varies when professor Alpha sits in various chairs. This will lead us to the unpleasant situation of counting unsystematically. The trick here is not to assign any professor a particular seat first. Instead, we arrange the four professors in relative positions and then assign them in their seats. By the given conditions, professors Alpha, Delta, and Gamma can sit in one of the following ways: (Alpha, Delta, Gamma), (Gamma, Alpha, Delta), (Delta, Alpha, Gamma), (Gamma, Delta, Alpha). For each of the above seating arrangements, professor Beta can sit on either end. Therefore, the answer is $2 + 2 + 2 + 2 = 8$. ■

> **Addition Principle.** If event A can occur in a ways and event B can occur in b other ways, then the event of either A or B can occur in $a + b$ ways.

This idea can easily be applied to more events. We can put the addition principle into the language of sets. Let S be a set. If $A_1, A_2, \ldots, A_n$ is a partition of S, then

$$|S| = |A_1| + |A_2| + \cdots + |A_n|,$$

where $|X|$ denotes the number of elements in the set X.

Example 1.3. Determine the number of squares with all their vertices belonging to the 10×10 by ten array of points defined in Figure 1.1. (The points are evenly spaced.)

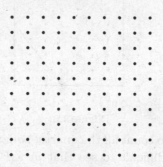

Figure 1.1.

Solution: We say that four of the points an $n \times n$ *quartet* if they are the vertices of an $n \times n$ square with its sides parallel to the borders of the array. We also say that a square with its vertices in a quartet is a *quartet square*. There are $9^2 = 81$ 1×1 quartets.

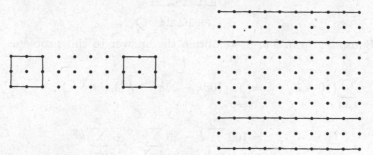

Figure 1.2.

It is easy to see that there are eight 2×2 quartets in the above 3×10 array of Figure 1.2. It is not difficult to see that there are eight such 3×10 arrays in the given 10×10 array. Hence there are 8^2 2×2 quartets. In exactly the same way, we can show that there are 7^2 3×3 quartets, and so on. That is, for $1 \le k \le 9$, there are $(10-k)^2$ $k \times k$ quartets. But the difficult part of this problem is that there are squares whose sides are not parallel to the borders of the given array. However, each of these squares is inscribed in a quartet square.

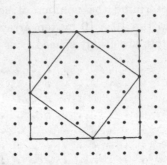

Figure 1.3.

Hence it suffices to count all the quartet squares and their inscribed squares. It is not difficult to see that in a k by k quartet square, there are k inscribed squares, including the quartet square itself. For example, for $k = 4$, we have Figure 1.4.

Figure 1.4.

Summing from 1 to 9, we obtain the answer to this problem:

$$\sum_{k=1}^{9}(10-k)^2 \cdot k = \sum_{k=1}^{9}(100k - 20k^2 + k^3)$$

$$= 100\sum_{k=1}^{9}k - 20\sum_{k=1}^{9}k^2 + \sum_{k=1}^{9}k^3$$

$$= 100 \cdot \frac{9 \cdot 10}{2} - 20 \cdot \frac{9 \cdot 10 \cdot 19}{6} + \left(\frac{9 \cdot 10}{2}\right)^2$$

$$= 4500 - 5700 + 2025 = 825,$$

by using the formulas

$$\sum_{k=1}^{n}k = \frac{n(n+1)}{2},$$

$$\sum_{k=1}^{n}k^2 = \frac{n(n+1)(2n+1)}{6},$$

and

$$\sum_{k=1}^{n} k^3 = \left(\frac{n(n+1)}{2}\right)^2.$$

∎

Example 1.4. There are n sticks of length $1, 2, \ldots, n$. How many incongruent triangles can be formed by using three of the given sticks?

Solution: Let x, y, z be the lengths of three sticks. Without loss of generality, we may assume that $x < y < z$. These three sticks can form a triangle if and only if x, y, z satisfy the **Triangle Inequality**; that is, $x + y > z$. We classify all incongruent triangles by their longest side. For $1 \leq k \leq n$, define

$$A_k = \{(x, y, z) \mid x, y, z \in \mathbb{Z},\ 1 \leq x < y < z = k,\ \text{and}\ x + y > z\}.$$

Hence, by the Addition Principle, we need to calculate

$$|A_1| + |A_2| + \cdots + |A_n|.$$

Under our assumption, $z \geq 3$. Hence $A_1 = A_2 = \emptyset$. If $z = 3$, then $x = 1$ and $y = 2$, and there is no triangle with side lengths 1, 2, 3. Thus $A_3 = \emptyset$. Hence $|A_1| = |A_2| = |A_3| = 0$. Now we assume that $k \geq 4$. We consider two cases.

- *Case 1.* In this case, we assume that k is even; that is, $k = 2m$ for some integer $m \geq 2$. Because $x < y$, $x + y > 2x$. Note also that $x + y > z$. We present different arguments for $2x \leq z$ and $2x > z$; that is, $1 \leq x \leq m$ and $m < x$.

  For $1 \leq x \leq m$, we need $y > z - x = k - x$. Since $k = 2m \geq 2x$, we know that $k - x \geq x$. Hence any y between $k - x + 1$ and $z - 1 = k - 1$, inclusive, will work, implying that there are $(k - 1) - (k - x + 1) + 1 = x - 1$ possible values for y.

  For $m < x$, the first inequality gives $x + y > 2x > 2m = z$. Thus any y between $x + 1$ and $k - 1$, inclusive, will work, implying that there are $(k - 1) - (x + 1) + 1 = k - x - 1 = 2m - x - 1$ possible values for y.

  Therefore, for $k = 2m$,

$$|A_k| = \sum_{x=1}^{m}(x-1) + \sum_{x=m+1}^{2m-1}(2m - x - 1) = \sum_{x=1}^{m}(x-1) + \sum_{i=0}^{m-2} i,$$

implying that

$$|A_k| = \frac{m(m-1)}{2} + \frac{(m-2)(m-1)}{2} = (m-1)^2. \qquad (\dagger)$$

Note that this formula also works for $m = 1$; that is, $k = 2$.

- *Case 2.* In this case, we assume that k is odd; that is, $k = 2m+1$ for some integer $m \geq 2$.

  For $1 \leq x \leq m$, we again need $y > z - x = k - x$. This time, $k = 2m + 1 > 2x$, so $k - x > x$. As before, y can take on all integer values between $k - x + 1$ and $k - 1$, inclusive, so there are $(k - 1) - (k - x + 1) + 1 = x - 1$ possible values for y.

  For $m < x$, identical reasoning as in the first case shows that any value of y such that $x < y < z$ will work. Thus there are $(k - 1) - (x - 1) + 1 = k - x - 1 = 2m - x$ possible values for y.

  Therefore, for $k = 2m + 1$,

$$|A_k| = \sum_{x=1}^{m}(x-1) + \sum_{x=m+1}^{2m}(2m-x) = \sum_{x=1}^{m}(x-1) + \sum_{i=0}^{m-1} i,$$

implying that

$$|A_k| = \frac{m(m-1)}{2} + \frac{m(m-1)}{2} = m(m-1). \qquad (**)$$

Note that this formula works also for $m = 0$ and $m = 1$; that is, $k = 1$ and $k = 3$.

Now we are ready to solve our problem.

If n is odd, then $n = 2p + 1$ for some nonnegative integer p. We have

$$|A_1| + |A_2| + \cdots + |A_n|$$
$$= (|A_1| + |A_3| + \cdots + |A_{2p+1}|) + (|A_2| + |A_4| + \cdots + |A_{2p}|)$$
$$= \sum_{m=0}^{p} m(m-1) + \sum_{m=1}^{p}(m-1)^2 = 2\sum_{m=1}^{p} m^2 - 3\sum_{m=0}^{p} m + p$$
$$= \frac{p(p+1)(2p+1)}{3} - \frac{3p(p+1)}{2} + p = p \cdot \frac{4p^2 + 6p + 2 - 9p - 9 + 6}{6}$$
$$= \frac{p(4p^2 - 3p - 1)}{6} = \frac{p(p-1)(4p+1)}{6}.$$

If n is even, then $n = 2p$ for some positive integer p. We have

$$|A_1| + |A_2| + \cdots + |A_n|$$

$$= (|A_1| + |A_3| + \cdots + |A_{2p-1}|) + (|A_2| + |A_4| + \cdots + |A_{2p}|)$$

$$= \sum_{m=0}^{p-1} m(m-1) + \sum_{m=1}^{p} (m-1)^2 = \sum_{m=1}^{p-1} m(m-1) + \sum_{m=1}^{p-1} m^2$$

$$= 2\sum_{m=1}^{p-1} m^2 - \sum_{m=1}^{p-1} m = \frac{(p-1)(p)(2p-1)}{3} - \frac{p(p-1)}{2}$$

$$= p(p-1) \cdot \frac{4p - 2 - 3}{6} = \frac{p(p-1)(4p-5)}{6}.$$

Putting the above together, we obtain

$$\begin{cases} \dfrac{p(p-1)(4p+1)}{6} & \text{triangles} \quad \text{if } n \text{ is odd and } n = 2p+1, \\[2mm] \dfrac{p(p-1)(4p-5)}{6} & \text{triangles} \quad \text{if } n \text{ is even and } n = 2p. \end{cases}$$

■

Once the problem was solved for odd $n = 2p+1$, the even case $n = 2p$ could be done just by calculating

$$|A_1| + |A_2| + \cdots + |A_{2p}|$$

$$= (|A_1| + |A_2| + \cdots + |A_{2p}| + |A_{2p+1}|) - |A_{2p+1}|$$

$$= \frac{p(p-1)(4p+1)}{6} - p(p-1)$$

$$= \frac{p(p-1)(4p-5)}{6}.$$

A more insightful approach to obtain equations (†) and (∗∗) uses lattice points; that is, points with integer coordinates in a coordinate plane. For $1 \le k \le n$, A_k contains the lattice points (x, y) that satisfy $x > 0$, $y > x$, $y < k$, and $x + y > k$. For example, for $k = 9$, we have Figure 1.5.

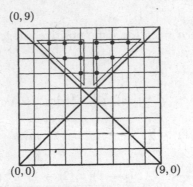

Figure 1.5.

Each triangle contains $1 + 2 + 3 = 6$ lattice points. This illustrates why $|A_k|$ is twice a triangular number when k is odd.

Similarly, for $k = 10$ we have Figure 1.6.

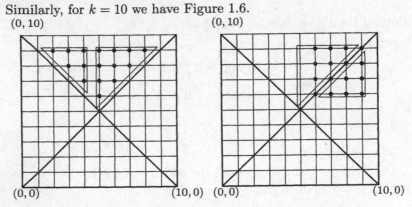

Figure 1.6.

If we move the left triangle containing 6 lattice points down and to the right of the right triangle containing 10 lattice points, we obtain a 4×4 array of lattice points. This illustrates why $|A_k| = (\frac{k}{2} - 1)^2$ when k is even.

Example 1.5. [PEA Math Materials, by Richard Parris] Before Rick can open his gym locker, he must remember the combination. Two of the numbers of the three-term sequence are 17 and 24, but he has forgotten the third, and does not know the order of the numbers. There are 40 possibilities for the third number. At ten seconds per try, at most how long will it take him to exhaust all possibilities?

Solution: We consider the six subsets of possible combinations. Let

$$A_1 = \{(x, 17, 24) \mid 1 \le x \le 40\},$$
$$A_2 = \{(x, 24, 17) \mid 1 \le x \le 40\},$$
$$A_3 = \{(17, x, 24) \mid 1 \le x \le 40\},$$
$$A_4 = \{(24, x, 17) \mid 1 \le x \le 40\},$$
$$A_5 = \{(17, 24, x) \mid 1 \le x \le 40\},$$
$$A_6 = \{(24, 17, x) \mid 1 \le x \le 40\}.$$

It is not difficult to see that each set has 40 elements. Thus by the addition principle there are $40 \cdot 6 = 240$ combinations to try and at most 40 minutes are needed. Not so fast, my friend! An important but easily missed fact in applying the addition principle is that the sets A_i must form a partition for the principle to hold; that is, $A_i \cap A_j = \emptyset$ for $i \ne j$. But in this problem, the combination $(17, 17, 24)$ belongs to both A_1 and A_3. Similarly, each of the combinations $(17, 24, 17)$, $(24, 17, 17)$, $(17, 24, 24)$, $(24, 17, 24)$, $(24, 24, 17)$ also belongs to two sets and is therefore counted twice. Hence there are only $240 - 6 = 234$ combinations to try, and the correct answer is 39 minutes. ∎

Addition and multiplication are closely related to each other. Multiplication is a shortcut for repeated addition. For example, $3 \cdot 5 = 3 + 3 + 3 + 3 + 3 = 5 + 5 + 5$. Using multiplication effectively can provide insight in solving counting problems. One might easily miss a double-counted combination in the last part of the solution of Example 1.5. One might also wonder whether there are more over-counted combinations. With a little bit more thought, we find that the only possible over-counted combinations are those consisting of numbers $\{a, a, b\}$ with $\{a, b\} = \{17, 24\}$. There are two possibilities for the values of a and b, namely, $(a, b) = (17, 24)$ and $(a, b) = (24, 17)$. There are three ways to arrange the numbers a, a, b, namely, (a, a, b), (a, b, a), and (b, a, a). Hence there are exactly 6 double-counted combinations.

We can also solve Example 1.2 using multiplication. First we arrange the relative positions of professors Alpha and Delta. There are two ways to do this, namely, (Alpha, Delta) and (Delta, Alpha). Professor Gamma has two choices to sit next to professors Alpha and Delta, namely, the two outside positions. Finally, professor Beta has two choices, namely, the two outside positions. Therefore, the answer is $2 \cdot 2 \cdot 2 = 8$. This discussion leads to another important counting principle.

> **Multiplication Principle.** If event A_1 can occur in a_1 different ways, and event A_2 can occur in a_2 different ways, ..., and event A_n can occur in a_n different ways, then the total number of ways that event A_1 followed by event A_2, ..., followed by event A_n can occur is $a_1 a_2 \cdots a_n$.

We can also put the multiplication principle into the language of sets; that is, if

$$S = \{(s_1, s_2, \ldots, s_n) \mid s_i \in S_i, 1 \le i \le n\},$$

then $|S| = |S_1| \cdot |S_2| \cdots |S_n|$.

Many combinatorial problems compute probabilities of certain events. The (classical) **probability** of an event E is the chance or likelihood that event E occurs.

Example 1.6. [AIME 1988] Compute the probability that a randomly chosen positive divisor of 10^{99} is an integer multiple of 10^{88}.

Solution: What are the divisors of 10^{99}? Is 3 a divisor? Is 220 a divisor? We consider the prime factorization of 10^{99}, or, $2^{99} \cdot 5^{99}$. The divisors of 10^{99} are of the form $2^a \cdot 5^b$, where a and b are integers with $0 \le a, b \le 99$. Since there are 100 choices for each of a and b, 10^{99} has $100 \cdot 100$ positive integer divisors. Of these, the multiples of $10^{88} = 2^{88} \cdot 5^{88}$ must satisfy the inequalities $88 \le a, b \le 99$. Thus there are 12 choices for each of a and b; that is, $12 \cdot 12$ of the $100 \cdot 100$ divisors of 10^{99} are multiples of 10^{88}. Consequently, the desired probability is $\frac{12 \cdot 12}{100 \cdot 100} = \frac{9}{625}$. ∎

Example 1.7. Determine the number of ordered pairs of positive integers (a, b) such that the least common multiple of a and b is $2^3 5^7 11^{13}$.

Solution: Both a and b are factors of $2^3 5^7 11^{13}$, and so $a = 2^x 5^y 11^z$ and $b = 2^s 5^t 11^u$ for some nonnegative integers x, y, z, s, t, u. Because $2^3 5^7 11^{13}$ is the least common multiple, $\max\{x, s\} = 3$, $\max\{y, t\} = 7$, and $\max\{z, u\} = 13$. Hence (x, s) can be equal to $(0, 3)$, $(1, 3)$, $(2, 3)$, $(3, 3)$, $(3, 2)$, $(3, 1)$, or $(3, 0)$, so there are 7 choices for (x, s). Similarly, there are 15 and 27 choices for (y, t) and (z, u), respectively. By the multiplication principle, there are $7 \times 15 \times 27 = 2835$ ordered pairs of positive integers (a, b) having $2^3 5^7 11^{13}$ as their least common multiple.   ∎

We can also use more multiplication in the previous solution. For example, to determine the number of ordered pairs (y, t), we can first choose one of y and t to be 7 for two possible choices, and then there are eight choices for the other variable, namely, $0, 1, 2, \ldots, 7$. Hence we obtain 2×8 possible choices with $(7, 7)$ being double counted, so there are $2 \times 8 - 1 = 15$ possible choices for (y, t).

Putting the last two examples together gives two interesting results in number theory. Let n be a positive integer. We can write $n = p_1^{a_1} p_2^{a_2} \cdots p_k^{a_k}$, where $p_1, p_2, \ldots, p_k$ are distinct primes and $a_1, a_2, \ldots, a_k$ are positive integers. By a basic result in number theory, this decomposition is unique up to rearrangements of the primes $p_1, p_2, \ldots, p_k$. This is called the **prime decomposition** of n.

Proposition 1.1.   *Let n be a positive integer, and let $n = p_1^{a_1} p_2^{a_2} \cdots p_k^{a_k}$ be a prime decomposition of n. Then n has*

$$(a_1 + 1)(a_2 + 1) \cdots (a_k + 1)$$

positive integer divisors, including 1 and itself.

Proposition 1.2.   *Let n be a positive integer, and let $n = p_1^{a_1} p_2^{a_2} \cdots p_k^{a_k}$ be a prime decomposition of n. Then there are*

$$(2a_1 + 1)(2a_2 + 1) \cdots (2a_k + 1)$$

distinct pairs of ordered positive integers (a, b) such that their least common multiple is equal to n.

The proofs of these two propositions are identical to those of Examples 1.6 and 1.7. It is interesting to note that these two results can be generalized to the case when the powers of the primes in the prime decomposition are nonnegative (because if $a_i = 0$ for some $1 \le i \le k$, then $a_i + 1 = 2a_i + 1 = 1$, which does not affect the products).

Example 1.8.   A license plate contains a sequence of three letters of the alphabet followed by a sequence of three digits. How many different license plates can be produced if 0 and O cannot be used at the same time?

Solution: Let S_1 denote the set of license plates with no 0's, and let S_2 denote the set of license plates with no O's. If $\alpha\beta\gamma - \theta\phi\psi$ is a plate in S_1, then $\theta, \phi, \psi \neq 0$. Consequently, there are no restrictions on α, β, γ; that is, for each of α, β, γ there are 26 choices, while for each of θ, ϕ, ψ there are nine choices. Therefore, $|S_1| = 26^3 \cdot 9^3$. In exactly the same way, $|S_2| = 25^3 \cdot 10^3$ (since the roles of letters and digits are switched). It seems that the answer to the problem is $|S_1| + |S_2| = 26^3 \cdot 9^3 + 25^3 \cdot 10^3$. However, this is not the correct answer. But every step seems logical. Where is the mistake? A more fundamental question is: How do we know whether there is a mistake?

We answer the second question first. Let S denote the set of all possible license plates containing a sequence of three letters followed by a sequence of three digits. Then there are 26 choices for each of the three letters and 10 choices for each of the digits. By the multiplication principle, $|S| = 26^3 \cdot 10^3$. It is not difficult to check that

$$|S_1| + |S_2| = 26^3 \cdot 9^3 + 25^3 \cdot 10^3 > 26^3 \cdot 10^3 = |S|.$$

This clearly shows that $|S_1| + |S_2|$ is not the desired answer. Now we are going to fix our mistake. We notice that there is some overlap in S_1 and S_2, namely, those license plates that have neither 0 nor O. Let S_3 denote the set of such plates. Then $S_3 = S_1 \cap S_2$. For each letter of a plate in S_3 there are 25 choices, and for each digit there are nine choices. Thus $|S_3| = 25^3 \cdot 9^3$. Since each plate in S_3 has been counted twice in S_1 and S_2, the final answer to our problem is

$$|S_1| + |S_2| - |S_3| = 26^3 \cdot 9^3 + 25^3 \cdot 10^3 - 25^3 \cdot 9^3 = 17047279.$$

∎

The technique of including overlapping sets and excluding the double-counted part is called the **Inclusion–Exclusion Principle**, which we will discuss carefully in Chapter 6.

Example 1.9. Determine the number of positive integers less than 1000 that contain at least one 1 in their decimal representation.

Solution: Let S be the set of all positive integers less than 1000; that is, $S = \{1, 2, \ldots, 999\}$ with $|S| = 999$. Let S_1, S_2, and S_3 denote the sets of positive integers with one digit, two digits, and three digits.

Then

$$S_1 = \{1, 2, \ldots, 9\},$$
$$S_2 = \{10, 11, \ldots, 99\},$$
$$S_3 = \{100, 101, \ldots, 999\}.$$

It is clear that S_1, S_2, S_3 is a partition of S, and $|S_1| = 9$, $|S_2| = 90$, and $|S_3| = 900$. For $i = 1, 2, 3$, let A_i be the subset of S_i containing exactly those numbers with at least one 1. It suffices to calculate

$$|A_1| + |A_2| + |A_3|.$$

It is easy to see that $A_1 = \{2, 3, \ldots, 9\}$ with $|A_1| = 8$. We can partition A_2 into three subsets:

$$A_2 = \{11\} \cup \{\overline{1b}, b \neq 1\} \cup \{\overline{a1}, a \neq 0 \text{ or } 1\}.$$

Then $|A_2| = 1 + 9 + 8 = 18$. Let $A_{\overline{1bc}}$ denote the set of three-digit numbers with leading digit 1 and $b, c \neq 1$, and let $A_{\overline{a1c}}$ denote the set of three-digit numbers with middle digit 1 and $a, c \neq 1$ and $a \neq 0$. We define other such sets analogously. We can partition A_3 as

$$A_3 = A_{\overline{111}} \cup A_{\overline{1bc}} \cup A_{\overline{a1c}} \cup A_{\overline{ab1}} \cup A_{\overline{11c}} \cup A_{\overline{1b1}} \cup A_{\overline{a11}}.$$

Then $A_{\overline{111}} = \{111\}$ and $|A_{\overline{111}}| = 1$. For $A_{\overline{1bc}}$, there are nine choices for each b and c, so $|A_{\overline{1bc}}| = 9 \cdot 9 = 81$. It is not difficult to see that $|A_{\overline{a1c}}| = |A_{\overline{ab1}}|$, because b and c are in symmetric positions. For $A_{\overline{a1c}}$, there are 8 choices for a and 9 choices for c, so $|A_{\overline{a1c}}| = |A_{\overline{ab1}}| = 8 \cdot 9 = 72$. It is easy to see that $|A_{\overline{11c}}| = |A_{\overline{1b1}}| = 9$ and $|A_{\overline{a11}}| = 8$. It follows that $|A_3| = 1 + 81 + 2 \cdot 72 + 2 \cdot 9 + 8 = 252$ and

$$|A_1| + |A_2| + |A_3| = 1 + 18 + 252 = 271.$$

■

We took a long way to solve this problem, which involves only addition and multiplication. However, we can put our solution on a diet with the assistance of subtraction.

We consider the following partition instead. Let S' be the set of nonnegative integers less than 1000, let B_1 be the set of nonnegative integers less than 1000 that contain at least one 1, and let B_2 be the set of nonnegative integers less than 1000 that do not contain 1. It is clear that $|S'| = 1000$, and that B_1, B_2 is a partition of S'. Hence

$$|B_1| + |B_2| = |S'| = 1000.$$

To solve the problem, we need to calculate $|B_1|$. Instead of thinking about the number of digits of a number, we allow 0's as leading digits. For example, 096 is simply 96. Hence every number in S' is now encoded by a three-digit code. Then B_2 contains those codes with no 1. There are 9 choices for the first place (leftmost place) of the code, namely, digits $0, 2, 3, \ldots, 9$. Similarly, there are 9 choices for each place in the code. Hence there are $9 \cdot 9 \cdot 9 = 9^3 = 729$ codes in B_2, and so $|B_1| = 1000 - 729 = 271$.

More importantly, if the condition 1000 is replaced by 10000, 100000, etc., our second method can be easily generalized. On the other hand, it would be extremely tedious to generalize our first method, because we need to discuss many more cases. For example, to consider four-digit numbers $\overline{abcd}$ with at least one 1, we need to consider $\overline{1bcd}$, $\overline{a1cd}$, $\overline{ab1d}$, $\overline{abc1}$, $\overline{11cd}$, $\overline{1b1d}$, $\overline{1bc1}$, $\overline{a11d}$, $\overline{a1c1}$, $\overline{ab11}$, $\overline{111d}$, $\overline{11c1}$, $\overline{1b11}$, $\overline{a111}$, $\overline{1111}$, for a total of 15 cases. Counting the number of cases itself becomes a good counting exercise! This new counting problem can be solved in a more insightful way by noting that $15 = 2^4 - 1$. Can you see why?

Example 1.10. There are 15 distinct air conditioning vents in a movie theater. To keep the air fresh, at least one of the vents has to be on at all times. In how many ways can this be done?

Solution: If we use the addition principle, there are many cases to consider. We label the vents $v_1, v_2, \ldots, v_{15}$. Each vent v_i has two choices: *on* or *off*. We use the letter n for "on" and the letter f for "off." We can assign a 15-letter code to each situation. For example,

$$nnfffnfnnffnnf$$

corresponds to the situation in which vents $v_1, v_2, v_7, v_9, v_{10}, v_{13}, v_{14}$ are on and all the other vents are off. By the multiplication principle, there are 2^{15} situations. But the code $fffffffffffffff$ is not allowed because it corresponds to the situation in which all vents are "off." It is clear that each situation matches exactly one code. Hence the answer is $2^{15} - 1 = 32767$.   ∎

In the last solution, we mapped each allowable situation to a unique code. We will discuss this **bijection** idea extensively later, in Chapter 4. If we replace 15 by n in the last problem, then we can prove, in exactly the same way, the following theorem.

Theorem 1.3. *Given a set S with $|S| = n$, there are 2^n subsets of S, including the empty set and S itself.*

Let A and B be two sets. A **map** (or **mapping** or **function**) f from A to B (written as $f : A \to B$) assigns to each $a \in A$ exactly one $b \in B$ (written $f(a) = b$); b is the **image** of a. For $A' \subseteq A$, let $f(A')$ (the image of A') denote the set of images of $a \in A'$. If $f(A) = B$, then f is **surjective** (or **onto**); that is, every $b \in B$ is the image of some $a \in A$. If every two distinct elements a_1 and a_2 in A have distinct images, then f is **injective** (or **one-to-one**). If f is both injective and surjective, then f is **bijective** (or a **bijection** or a **one-to-one correspondence**). A **permutation** is a change in position within a collection. More precisely, if S is a set, then a permutation of S is a one-to-one function π that maps S onto itself. If $S = \{x_1, x_2, \ldots, x_n\}$ is a finite set, then we may denote a permutation π of S by $(y_1, y_2, \ldots, y_n)$, where $y_k = \pi(x_k)$. A straightforward use of the multiplication principle proves the following theorem.

Theorem 1.4. *Let n and k be positive integers with $n \geq k$. The total number of permutations of n objects taken k at a time is*

$$_nP_k = n(n-1)\cdots(n-k+1) = \frac{n!}{(n-k)!},$$

where $0! = 1$ and $n! = 1 \cdot 2 \cdots\cdot n$, for $n \geq 1$.

Setting $k = n$ in Theorem 1.4, we have the following corollary.

Corollary 1.5. *The total number of permutations of n objects is $n!$.*

Example 1.11. [AIME 1996] In a five-team tournament, each team plays one game with every other team. Each team has a 50% chance of winning any game it plays. (There are no ties.) Compute the probability that the tournament will produce neither an undefeated team nor a winless team.

Solution: Each team has to play four games. Hence there are $5 \cdot 4$ games if one counts each game twice, once by each of the two teams playing the game. The five teams play a total of $\frac{5 \cdot 4}{2} = 10$ games.

Because each game can have two possible outcomes, by Theorem 1.3, there are 2^{10} possible outcomes for the tournament.

There are five ways to choose an undefeated team. Say team A wins all four of its games. Then each of the remaining six games has two possible outcomes for a total of $2^{10-4} = 2^6$ outcomes. Because at most one team can be undefeated, there are $5 \cdot 2^6$ tournaments that produce an undefeated team. A similar argument shows that $5 \cdot 2^6$ of the 2^{10} possible tournaments produce a winless team.

However, these possibilities are not mutually exclusive. It is possible to have exactly one undefeated team and exactly one winless team in the same tournament. There are $_5P_2 = 5 \cdot 4 = 20$ such two-team permutations. Say team A is undefeated and team B is winless. There are seven (not eight, because A and B play against each other!) games in which either team A or team B or both teams play. The outcomes of these seven games are decided. Each of the remaining three games left has two possible outcomes for a total of $2^{10-7} = 2^3$ tournaments. In other words, $20 \cdot 2^3 = 5 \cdot 2^5$ of the 2^{10} tournaments have both an undefeated team and a winless team. Thus, according to the Inclusion–Exclusion principle, there are

$$2^{10} - 2 \cdot 5 \cdot 2^6 + 5 \cdot 2^5 = 2^5(2^5 - 5 \cdot 2^2 + 5) = 2^5 \cdot 17$$

tournament outcomes in which there is neither an undefeated nor a winless team. All outcomes are equally likely; hence the required probability is $\frac{17 \cdot 2^5}{2^{10}} = \frac{17}{32}$. ∎

Example 1.12. Determine the number of seven-letter codes such that

(i) no letters are repeated in the code; and

(ii) letters a and b are not next to each other.

First Solution: By Theorem 1.2, there are $_{26}P_7$ distinct codes satisfying condition (i) only. Let S denote the set of all such codes. We partition S into three sets:

$$S_1 = \{s \in S \mid s = \ldots ab \ldots\},$$
$$S_2 = \{s \in S \mid s = \ldots ba \ldots\},$$
$$S_3 = S \setminus S_1 \setminus S_2.$$

It is clear that S_3 consists of all the codes satisfying the conditions of the problem, and it suffices to calculate $|S_3|$. Note that each code in S_1 can be mapped uniquely to a code in S_2 by switching ab to ba, and vice versa, so $|S_1| = |S_2|$. To generate a code in S_1, we first select 5 letters $\alpha_1, \alpha_2, \ldots, \alpha_5$ in order from the 24-letter set $\{c, d, \ldots, z\}$ to form the code

$$\_\alpha_1\_\alpha_2\_\alpha_3\_\alpha_4\_\alpha_5\_.$$

Then we note that there are 6 underlined spaces in which to place ab. Hence $|S_1| = |S_2| =_{24} P_5 \cdot 6$, and

$$|S_3| =_{26} P_7 - 2 \cdot_{24} P_5 \cdot 6 = 24 \cdot 23 \cdot 22 \cdot 21 \cdot 20(26 \cdot 25 - 12) = 3254106240.$$

Second Solution: As in the first solution, S denotes the set of all seven-letter codes with no repeating letters. Let

$A_1 = \{s \in S \mid s \text{ contains neither } a \text{ nor} b\},$

$A_2 = \{s \in S \mid s \text{ contains } a \text{ but not } b\},$

$A_3 = \{s \in S \mid s \text{ contains } b \text{ but not } a\},$

$A_4 = \{s \in S \mid \mid s \text{ contains } a \text{ and } b, \text{ not next to each other}\}.$

It suffices to calculate $|A_1| + |A_2| + |A_3| + |A_4|$. By Theorem 1.2, $|A_1| =_{24} P_7$. As in the first solution, we can show that $|A_2| = |A_3|$. To compute $|A_2|$, we construct the elements in A_2 in two steps. We first pick 6 letters $\alpha_1, \alpha_2, \ldots, \alpha_6$ in order from the 24-letter set $\{c, d, \ldots, z\}$ to form the code

$$\_\alpha_1\_\alpha_2\_\alpha_3\_\alpha_4\_\alpha_5\_\alpha_6'\_.$$

Then we pick an empty space for a. Hence $|A_2| = |A_3| =_{24} P_6 \cdot 7$. To calculate $|A_4|$, we construct the elements in A_4 in two steps. We first pick 5 letters $\alpha_1, \alpha_2, \ldots, \alpha_5$ in order from $\{c, d, \ldots, z\}$ to form the code

$$\_\alpha_1\_\alpha_2\_\alpha_3\_\alpha_4\_\alpha_5\_.$$

Then we pick an empty space for a and another one for b. Hence $|A_4| =_{24} P_5 \cdot 6 \cdot 5$. Thus the answer to the problem is

$$|A_1| + |A_2| + |A_3| + |A_4| = {}_{24}P_7 + 2 \cdot_{24} P_6 \cdot 7 +_{24} P_5 \cdot 6 \cdot 5$$

$$= 3254106240.$$

■

Example 1.13. Dr. A celebrated his 24th birthday on 19 August 1980. He noticed that exactly eight years earlier there was a $19 \sim 8$ date yielding a number divisible by 198 and not divisible by 1980. A calendar date $d_1 d_2 / m_1 m_2 / y_1 y_2$ (day-month-year) is called $19 \sim 8$ if $d_1 + m_1 + y_1 = 8$ and $d_2 + m_2 + y_2 = 19$. For how many $19 \sim 8$ dates is the corresponding six-digit number $\overline{d_1 d_2 m_1 m_2 y_1 y_2}$ (leading zero allowed) divisible by 198 and not divisible by 1980?

To solve this problem, we need a little bit knowledge of number theory. Let $n = \overline{a_m a_{m-1} \ldots a_1 a_0}$ be an $(m+1)$-digit number (in base 10). Then

$$n = a_m \cdot 10^m + a_{m-1} \cdot 10^{m-1} + \cdots + a_1 \cdot 10 + a_0.$$

For any nonnegative k, 10^k has remainder 1 when divided by 9; that is, $10^k \equiv 1 \pmod 9$. Thus the remainder when n is divided by 9 is equal to that of the sum of the digits of n divided by 9; that is,

$$\overline{a_m a_{m-1} \ldots a_1 a_0} \equiv a_m + a_{m-1} + \cdots + a_0 \pmod 9.$$

For any nonnegative k, $10^{2k} \equiv 1 \pmod{11}$ and $10^{2k+1} \equiv -1 \pmod{11}$. Thus the remainder when n divided by 11 is equal to that of the alternating sum of the digits of n divided by 11; that is,

$$\overline{a_m a_{m-1} \ldots a_1 a_0} \equiv (-1)^m a_m + (-1)^{m-1} a_{m-1} + \cdots - a_1 + a_0 \pmod{11}.$$

Solution: The sum of the digits of the number $c = \overline{d_1 d_2 m_1 m_2 y_1 y_2}$ corresponding to a $19 \sim 8$ date is $d_1 + d_2 + m_1 + m_2 + y_1 + y_2 = (d_1 + m_1 + y_1) + (d_2 + m_2 + y_2) = 8 + 19 = 27$, which is divisible by 9, so c is divisible by 9. The alternating sum of this number's digits is $d_2 - d_1 + m_2 - m_1 + y_2 - y_1 = (d_2 + m_2 + y_2) - (d_1 + m_1 + y_1) = 19 - 8 = 11$, so c is divisible by 11. Hence any $19 \sim 8$ calendar date yields a six-digit number that is divisible by 99. It follows that this number is divisible by 198 but not divisible by 1980 if and only if its last digit is even and different from 0. By the Multiplication Principle, the answer seems to be mn, where m is the number of ordered triples of digits (d_1, m_1, y_1) such that $d_1 + m_1 + y_1 = 8$, and n is the number of ordered triples of digits (d_2, m_2, y_2) such that $d_2 + m_2 + y_2 = 19$, and $y_2 \in \{2, 4, 6, 8\}$.

But not so fast, my friend! Will all of these solutions yield valid dates? Consider, for example, the solution $(d_1, m_1, y_1) = (3, 4, 1)$, $(d_2, m_2, y_2) = (8, 9, 2)$. Although these ordered pairs certainly satisfy the equations, 38/49/12 is not a date. Thus we must be a little more careful. More specifically, we must ensure that $0 < \overline{m_1 m_2} \leq 12$ and $0 < \overline{d_1 d_2} \leq 28, 29, 30$, or 31, depending on the month and the year.

First note that because $y_2 \leq 8$ and $d_2 + m_2 + y_2 = 19$, $d_2 \geq 2$. Thus, since $\overline{d_1 d_2} < 32$ always, the only choices for d_1 are 0, 1, and 2. Furthermore (except possibly when $\overline{m_1 m_2} = 02$), all three choices will always work. Next, consider the possibilities for m_1. Clearly, $m_1 = 0$ or 1. By reasoning similar to that above, we also have $m_2 \geq 2$, so that $m_1 = 0$ will always work and $m_1 = 1$ will only work when $m_2 = 2$. Putting these facts together (and knowing that every choice of (d_1, m_1) will determine a unique possibility for y_1), we conclude that a given triplet (d_2, m_2, y_2) will yield $3 \cdot 1 = 3$ valid possibilities for (d_1, m_1, y_1) unless $m_2 = 2$, when there are $3 \cdot 2 = 6$ valid dates.

Now we are ready to count the triples (d_2, m_2, y_2). When $y_2 = 2$, we have $(d_2, m_2) = (8, 9), (9, 8)$. For $y_2 = 4$, we obtain $(d_2, m_2) = (6, 9), (7, 8), (8, 7), (9, 6)$. For $y_2 = 6$, we have $(d_2, m_2) = (4, 9), (5, 8), (6, 7), (7, 6), (8, 5), (9, 4)$. For $y_2 = 8$, the possibilities are $(d_2, m_2) = (2, 9), (3, 8), (4, 7), (5, 6), (6, 5), (7, 4), (8, 3), (9, 2)$. In all, we have 19 triples (d_2, m_2, y_2) with $m_2 \neq 2$ and one triple $(9, 2, 8)$ with $m_2 = 2$. Checking the case $\overline{m_1 m_2} = 02, d_1 = 2$, we see that although $\overline{d_1 d_2} = 29$, we have $\overline{y_1 y_2} = 68 \equiv 0 \pmod 4$, so that this is indeed a date.

Thus the answer to the problem is $19 \cdot 3 + 1 \cdot 6 = 63$. ∎

A **circular permutation** is a permutation of objects arranged in a circular fashion. Two such permutations are considered the same if one can be obtained from the other by rotation.

Corollary 1.6. *The total number of different circular permutations for n objects is $(n-1)!$.*

Example 1.14. Claudia has cans of paint in eight different colors. She wants to paint the four unit squares of a 2×2 board in such a way that neighboring unit squares are painted in different colors.

Determine the number of distinct coloring schemes Claudia can make. Two coloring schemes are considered the same if one can be obtained from the other by rotation.

Solution: Claudia needs at least two and at most four colors. There are three cases shown in Figure 1.7.

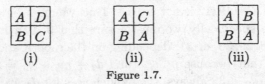

Figure 1.7.

In case (i), there are $_8P_4$ ways to choose distinct colors A, B, C, and D. Each coloring scheme in this case can be rotated 90 degrees counterclockwise three times to obtain three different coloring schemes as shown in Figure 1.8. In other words, each coloring scheme in this case is counted four times, considering rotations. Hence there are $\frac{_8P_4}{4} = \frac{8 \cdot 7 \cdot 6 \cdot 5}{4} = 420$ different coloring schemes.

A	D		D	C		C	B		B	A
B	C		A	B		D	A		C	D

Figure 1.8.

In case (ii), there are $_8P_3$ ways to choose distinct colors A, B, and C. Each coloring scheme in this case can be rotated 90 degrees counterclockwise three times to obtain three different coloring schemes as shown in Figure 1.9. In other words, each coloring scheme in this case is counted four times, considering rotations. Hence there are $\frac{_8P_3}{4} = \frac{8 \cdot 7 \cdot 6}{4} = 84$ different coloring schemes.

A	C		C	A		A	B		B	A
B	A		A	B		C	A		A	C

Figure 1.9.

In case (iii), there are $_8P_2$ ways to choose distinct colors A and B. Each coloring scheme in this case can be rotated 90 degrees counterclockwise once to obtain another different coloring scheme as shown in Figure 1.10. Each coloring scheme in this case is counted twice, considering rotations. Hence there are $\frac{_8P_2}{2} = \frac{8 \cdot 7}{2} = 28$ different coloring schemes.

Figure 1.10.

Consequently, there are $420 + 84 + 28 = 532$ distinct coloring schemes. ∎

Are we done? Not so fast my friend! The reader might have already obtained a different answer. But before we point out our mistake, we might want to ask how to detect a possible mistake in counting. Well, one effective way is to apply the identical method to different initial values. In this example, the number of colors given is certainly not crucial in our solution. What if we were given seven colors initially? Well, then we would find $\frac{7P_3}{4} = \frac{105}{2}$ distinct coloring schemes in case (ii). Indeed, we do not have four distinct schemes in Figure 1.9. The third scheme from the left is the same as the the first scheme, because the distributions of colors B and C are counted in choosing colors in order $(_8P_3)$. Likewise, the second and the fourth schemes are the same when choosing colors in order. Thus, there are $\frac{_8P_3}{2} = 168$ distinct schemes in case (ii). Hence, the correct answer for Example 1.14 is $420 + 168 + 28 = 616$.

Exercises 1

1.1. Determine the number of functions $f : \{1, 2, \ldots, 1999\} \longrightarrow \{2000, 2001, 2002, 2003\}$ satisfying the condition that $f(1)+f(2)+ \cdots + f(1999)$ is odd.

1.2. Find the number of two-digit positive integers that are divisible by both of their digits.

1.3. In how many ways can COMPUTER be spelled by moving either down or diagonally to the right in Figure 1.11.?

```
C
O  O
M  M  M
P  P  P  P
U  U  U  U  U
T  T  T  T  T  T
E  E  E  E  E  E  E
R  R  R  R  R  R  R  R
```
Figure 1.11.

1.4. [AHSME 1991] For any set S, let $|S|$ denote the number of elements in S, and let $n(S)$ be the number of subsets of S, including the empty set and S itself. If A, B, and C are sets for which

$$n(A) + n(B) + n(C) = n(A \cup B \cup C) \quad \text{and} \quad |A| = |B| = 100,$$

then what is the minimum possible value of $|A \cap B \cap C|$?

1.5. The baseball team PEA Twins consists of twelve pairs of twin brothers. On the first day of spring training, all 24 players stand in a circle in such a way that all pairs of twin brothers are neighbors. In how many ways can this be done? (Hint: look at Figure 1.12.)

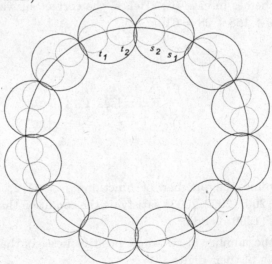

Figure 1.12.

1.6. [AIME 1992] For how many pairs of consecutive integers in

$\{1000, 1001, 1002, \ldots, 2000\}$ is no carrying required when the two integers are added?

1.7. [AIME 2000] Each of two boxes contains both black and white marbles, and the total number of marbles in the two boxes is 25. One marble is taken out of each box randomly. The probability that both marbles are black is $\frac{27}{50}$. What is the probability that both marbles are white?

1.8. There are ten girls and four boys in Mr. Fat's combinatorics class. In how many ways can these students sit around a circular table such that no boys are next to each other?

1.9. Find the number of ordered triples of sets (A, B, C) such that $A \cup B \cup C = \{1, 2, \ldots, 2003\}$ and $A \cap B \cap C = \emptyset$.

1.10. [ARML 2001] Compute the number of sets of three distinct elements that can be chosen from the set $\{2^1, 2^2, 2^3, \ldots, 2^{2000}\}$ such that the three elements form an increasing geometric progression.

1.11. Let n be an integer greater than four, and let $P_1 P_2 \ldots P_n$ be a convex n-sided polygon. Zachary wants to draw $n - 3$ diagonals that partition the region enclosed by the polygon into $n - 2$ triangular regions and that may intersect only at the vertices of the polygon. In addition, he wants each triangular region to have at least one side that is also a side of the polygon. In how many different ways can Zachary do this?

1.12. How many five-digit numbers are divisible by three and also contain 6 as one of their digits?

2
Combinations

We often deal with problems requiring us to count the number of **combinations**; that is, sets of unordered objects.

Example 2.1. Coach Z has reduced his roster to 12 players for this year's basketball playoff season. Among these 12 players, six can play either center or forward and seven can play guard, and the rookie Slash can play all three positions. How many different one center/two forwards/two guards lineups can coach Z have? (Lineups with the same player playing at different positions are counted as different lineups.)

Solution: We consider three cases.

- *Case 1.* In this case, we assume that Slash plays center. There are five players left for the forward position. There are $_5P_2 = 5 \times 4 = 20$ ways to pick two of them in order. But picking forward A first and then picking forward B second gives the same forward-lineup as picking B first and A second. Each of the forward lineups has been counted twice. Hence there are $20/2 = 10$ forward lineups. Likewise, there are six players to play guard. There are $_6P_2 = 6 \cdot 5 = 30$ ways to pick two of them in order. Again, since the order does not matter, we have $30/2 = 15$ guard lineups. Hence there are $10 \cdot 15 = 150$ lineups in this case.

- *Case 2.* In this case, we assume that Slash plays forward. There are five players left for the center and forward positions. There are five choices for the center, and so there are four choices left for another forward. As in Case 1, there are 15 guard lineups. Hence there are $5 \cdot 4 \cdot 15 = 300$ lineups in this case.

- *Case 3.* In this case, we assume that Slash plays guard. There are six players left who can play guard, so there are six choices for the other guard position in the lineup. There are five players left for the center and forward positions. There are five choices for the center. Among the four remaining players, there are $_4P_2/2 = 6$ forward lineups. (The reason for dividing by 2 is the same as in Case 1.) Hence there are $5 \cdot 6 \cdot 6 = 180$ lineups in this case.

- *Case 4.* In this case, we assume that Slash does not play. There are five players for the center and forward positions. As in Case 3, there are $5 \cdot 6 = 30$ center–forward lineups. There are six players playing guard. As in Case 1, there are 15 guard lineups. Hence there are $30 \cdot 15 = 450$ lineups in this case.

Putting the above together, we have a total of $150 + 300 + 180 + 450 = 1080$ different lineups for coach Z. ∎

Example 2.2. [AIME 1983] Twenty-five of King Arthur's knights are seated at their customary round table. Three of them are chosen–all choices of three being equally likely–and are sent off to slay a troublesome dragon. What is the probability that at least two of the three were seated next to each other?

First Solution: It is perhaps easier to think in terms of n knights, where $n \geq 4$.

We first count the ways of selecting three knights if there are no restrictions. We can use the multiplication principle. We have n ways to choose the first knight, then $n-1$ ways to choose the second knight, then $n-2$ ways to choose the third knight. But the order of choosing these three knights to form the triplet does not matter. A triplet A, B, C can be chosen in $3! = 6$ ways, namely, ABC, ACB, BAC, BCA, CAB, CBA. Therefore, each triplet is repeated six times in the selection procedure. Thus there are $T_n = \frac{n(n-1)(n-2)}{6}$ ways to choose triplets without any restrictions. Let S_n be the number of those triplets that include at least two neighbors, and let P_n be the probability that at least two of the three knights were neighbors. Then $P_n = \frac{S_n}{T_n}$.

To compute S_n, we need to use both the addition principle and the multiplication principle. We begin by breaking the selection procedure into two cases. Case 1 is straightforward. In Case 2, however, we need to complete our selection procedure into two steps.

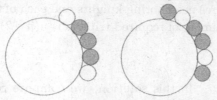

Figure 2.1.

- *Case 1.* The three knights are neighbors. Considering each knight along with the two knights to his immediate right, we see that there are n ways to pick three neighboring knights.

- *Case 2.* Exactly two of the three knights are neighbors. There are n ways to pick two neighboring knights (as with three) followed by $n-4$ ways of picking a third non-neighboring knight. (We must avoid the pair and the two knights on either side; otherwise, all three knights are neighbors.) Thus, there are $n(n-4)$ triplets that include exactly two neighbors.

Putting cases 1 and 2 together, there are $n + n(n-4) = n(n-3)$ ways to have at least two of the three knights sitting together; that is, $S_n = n(n-3)$. It follows that

$$P_n = \frac{n(n-3)}{\frac{n(n-1)(n-2)}{6}} = \frac{6(n-3)}{(n-1)(n-2)}.$$

Then the desired probability is $P_{25} = 11/46$. ∎

Most counting problems can be solved in various ways. But one always needs to analyze cases carefully. We present two other methods of computing S_n.

Second Solution: There are n ways to pick two neighboring knights followed by $n-2$ ways of picking a third knight. Here our situation is different from that of Case 2 because we do not need to avoid both the pair and the two knights on either side. Thus it is tempting to say that there are $n(n-2)$ ways that count both the situations of three neighboring knights and exactly two neighboring knights. But each selection containing three neighbors is counted twice. Let A, B, C be three neighboring knights sitting in that order. The selection of A, B, C is counted first by choosing C as the third knight to the

neighboring pair A and B, and then counted again by choosing A as the third knight to the neighboring pair B and C. Since there are n ways to choose three neighboring knights and each of them is counted twice, the total number of required triplets is $n(n-2)-n = n(n-3)$.
∎

Third Solution: In this solution, we compute S_n by using the multiplication principle and by moving in clockwise direction. We consider the first (clockwise) neighboring pair in the triplet. This pair is uniquely determined. Again there are n ways to pick two neighboring knights as this pair, because we are moving in a clockwise direction. In the second solution we found that there were $n-2$ ways of picking the third knight. But in this solution there are only $n-3$ ways of picking the third knight. This is because we cannot pick the knight right before the pair in clockwise order; otherwise, the neighboring pair we picked is not the first clockwise pair in the triplet. Therefore, we again have $n(n-3)$ triplets satisfying the required conditions.   ∎

Theorem 2.1.   *Let n and k be positive integers with $n \geq k$. The total number of combinations of n objects taken k at a time is*

$$\binom{n}{k} = \frac{n(n-1)\cdots(n-k+1)}{k!} = \frac{n!}{(n-k)!k!}.$$

It is also common to use $_nC_k$ or $C(n,k)$ instead of $\binom{n}{k}$.

Proof: There are $_nP_k = \frac{n!}{(n-k)!}$ ways to pick k objects in order. There are $k!$ ways to arrange the picked object; that is, there are $k!$ ways to pick the same k-object combination. Thus each combination is counted $k!$ times in $_nP_k$. Hence there are $\frac{n!}{(n-k)!k!}$ combinations. ∎

In situations in which the order in which the objects are to be arranged is somehow predetermined, the order of picking the objects is no longer important.

Example 2.3.   In Mr. Fat's math class, there are five boys and nine girls. At the end of the term, Mr. Fat wants to take a picture of the whole class. He wants all students to stand in a row, with boys standing in decreasing order according to their height (assuming that they have distinct heights) from left to right and girls standing in increasing order according to their height (assuming that they have distinct heights) from left to right. In how many ways can this be done? (The boys need not stand together, and the girls need not stand together.)

Solution: Consider 14 empty spaces in a row

$$s_1 s_2 \ldots s_{14}.$$

If $s_{i_1}, s_{i_2}, \ldots, s_{i_5}$ are picked for the boys' positions, then the boys are to be put in those places in a unique way according to their heights. The girls then also have a unique way to place themselves. Hence the answer is

$$\binom{14}{5} = \frac{14!}{5! \cdot 9!} = \frac{14 \cdot 13 \cdot 12 \cdot 11 \cdot 10}{1 \cdot 2 \cdot 3 \cdot 4 \cdot 5} = 14 \cdot 13 \cdot 11 = 2(7 \cdot 11 \cdot 13) = 2002.$$

∎

Example 2.4. Three distinct vertices are randomly chosen among the vertices of a cube. What is the probability that they are the vertices of an equilateral triangle? (See Figure 2.2.)

Solution:

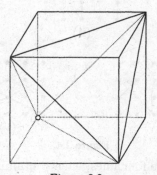

Figure 2.2.

The three vertices may be chosen among the eight vertices of the cube in $\binom{8}{3} = 56$ ways. Now, in order to have an equilateral triangle, the sides must be face diagonals. There are three face diagonals coming out of each vertex. Any two of such three face diagonals are the sides of an equilateral triangle. Hence each vertex is the vertex of $\binom{3}{2} = 3$ distinct equilateral triangles. Your initial response might be that there are 8×3 equilateral triangles. Note, however, that each equilateral triangle is counted three times in this way, because it is counted once for each of its vertices. Thus there are $\frac{24}{3} = 8$ distinct equilateral triangles. Thus the answer to the problem is $\frac{8}{56} = \frac{1}{7}$. ∎

Example 2.5. Figure 2.3 shows a 4×5 array of points, each of which is 1 unit away from its nearest neighbors. Determine the number of non-degenerate triangles (i.e., triangles with positive area) whose vertices are points in the given array.

Figure 2.3.

Solution: There are $\binom{20}{3} = \frac{20\cdot19\cdot18}{3!} = 1140$ ways to choose three points. But not all of these triplets form non-degenerate triangles. We need to find how many triplets are collinear. We consider the five cases shown in Figure 2.4.

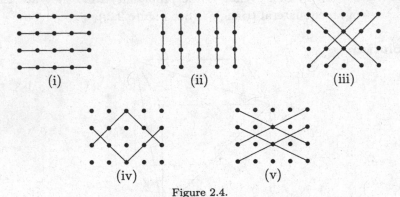

Figure 2.4.

In case (i), there are four lines, each containing five points. Hence there are $4 \times \binom{5}{3} = 40$ collinear triplets. In case (ii) and case (iii) combined, there are nine lines altogether, each containing four points. Hence there are $9 \times \binom{4}{3} = 36$ collinear triplets. In case (iv) and case (v) combined, there are eight lines altogether, each containing three points. Thus there are $8 \times \binom{3}{3} = 8$ collinear triplets.

Thus the answer is $1140 - (40 + 36 + 8) = 1056$. ∎

Example 2.6. There are 210 boys attending the United Football Summer Camp. Each of them is assigned to work with one of the twenty coaches. It is noted that each coach works with a distinct number of boys. In how many ways can the groups of boys be assigned?

Solution: Note that $1+2+\cdots+20 = \frac{20\cdot 21}{2} = 210$. For each number between 1 and 20 inclusive, there is a coach working with exactly that number of boys. Hence there are

$$\binom{210}{1}\binom{209}{2}\binom{207}{3}\cdots\binom{39}{19}\binom{20}{20}$$

$$= \frac{210!}{1!\cdot 209!}\cdot\frac{209!}{2!\cdot 207!}\cdot\frac{207!}{3!\cdot 204!}\cdots\frac{39!}{19!\cdot 20!}\cdot\frac{20!}{20!\cdot 0!}$$

$$= \frac{210!}{1!2!3!\cdots 20!}$$

possible different groups. ■

Corollary 2.2. *Let* m, n, $k_1, k_2, \ldots, k_m$ *be positive integers with* $n \geq k_1 + k_2 + \cdots + k_m$. *The total number of combinations of* n *objects taken* $k_1, k_2, \ldots, k_m$ *at a time, in that order, is*

$$\binom{n}{k_1, k_2, \ldots, k_m, k_{m+1}} = \frac{n!}{k_1!k_2!\cdots k_m!k_{m+1}!},$$

where $k_{m+1} = n - (k_1 + k_2 + \cdots + k_m)$.

Proof: By Theorem 2.1, there are $\binom{n}{k_1}$ ways to take k_1 objects at a time. There are $n - k_1$ objects left. By Theorem 2.1 again, there are $\binom{n-k_1}{k_2}$ ways to pick another k_2 objects at a time. By repeating this process, we conclude that the number of combinations of n objects taken $k_1, k_2, \ldots, k_m$ at a time, in that order, is

$$\binom{n}{k_1}\binom{n-k_1}{k_2}\binom{n-k_1-k_2}{k_3}\cdots\binom{n-k_1-k_2-\cdots-k_{m-1}}{k_m}$$

$$= \frac{n!}{k_1!(n-k_1)!}\cdot\frac{(n-k_1)!}{k_2!(n-k_1-k_2)!}\cdot\frac{(n-k_1-k_2)!}{k_3!(n-k_1-k_2-k_3)!}$$

$$\cdots\frac{(n-k_1-k_2-\cdots-k_{m-1})!}{k_m!(n-k_1-k_2-\cdots-k_m)!}$$

$$= \frac{n!}{k_1!k_2!\cdots k_m!k_{m+1}!},$$

as desired. ■

Example 2.7. [AIME 1990] In a shooting match, eight clay targets are arranged in two hanging columns of three each and one column of two (Figure 2.5). A marksman is to break all eight targets according to the following rules: (1) The marksman first chooses a column from which a target is to be broken. (2) The marksman must then break the lowest remaining unbroken target in the chosen column. If these rules are followed, in how many different orders can the eight targets be broken?

Figure 2.5.

Solution: Consider the eight shots that must be fired to break the eight targets. Of the eight, any subset of three shots may be the shots used to break the targets in the first column (but once these three shots are chosen, the rules of the match determine the order in which the targets will be broken.) This set of shots for the first column may be chosen in $\binom{8}{3}$ ways. From the remaining five shots, the three used to break the targets in the other column of three may be chosen in $\binom{5}{3}$ ways, while the remaining two shots will be used to break the remaining two targets. Combining, we find that the number of orders in which the targets can be broken is

$$\binom{8}{3}\binom{5}{3}\binom{2}{2} = \binom{8}{3,3,2} = \frac{8!}{3!3!2!} = 560.$$

■

Example 2.8. The PEA mathematics department is to hold a meeting to discuss pedagogy. After a long conversation among 23 members of the department, they decide to split into 5 groups of three and 2 groups of four to continue their discussion. In how many ways can this be done?

Solution: It seems that there are

$$\binom{23}{3,3,3,3,3,4,4} = \frac{23!}{(3!)^5(4!)^2}$$

ways to group them. But not so fast, my friend! Since each group is working on the same subject, there should be no ordering of the groups of the same size. Hence the answer is

$$\frac{1}{2!5!}\binom{23}{3,3,3,3,3,4,4} = \frac{23!}{2!(3!)^5(4!)^25!}.$$

∎

In solving many problems that ask for the probability of a certain event, one can choose freely whether the objects are distinguishable or not. In other words, one can choose to use $_nP_k$ or $\binom{n}{k}$, as long as one is consistent.

Example 2.9. [AIME 1984] A gardener plants three maple trees, four oak trees, and five birch trees in a row. He plants them in random order, each arrangement being equally likely. What is the probability that no two birch trees are next to one another?

First Solution: The 12 trees can be planted in 12! ways. Let k be the number of ways in which no two birch trees are adjacent to one another. The probability we need is $k/(12!)$. To find k, we will count the number of patterns shown in Figure 2.6,

$$\underset{1}{\rule{0pt}{0pt}} \overset{N}{\rule{1em}{0.4pt}} \underset{2}{\rule{0pt}{0pt}} \overset{N}{\rule{1em}{0.4pt}} \underset{3}{\rule{0pt}{0pt}} \overset{N}{\rule{1em}{0.4pt}} \underset{4}{\rule{0pt}{0pt}} \overset{N}{\rule{1em}{0.4pt}} \underset{5}{\rule{0pt}{0pt}} \overset{N}{\rule{1em}{0.4pt}} \underset{6}{\rule{0pt}{0pt}} \overset{N}{\rule{1em}{0.4pt}} \underset{7}{\rule{0pt}{0pt}} \overset{N}{\rule{1em}{0.4pt}} \underset{8}{\rule{0pt}{0pt}}$$

Figure 2.6.

where the seven N's denote non-birch (i.e., maple and oak) trees, and slots 1 through 8 are occupied by birch trees, with at most one in each slot. There are 7! orders for the non-birch trees, and for each ordering there are $_8P_5 = \frac{8!}{3!}$ ways to place the birch trees. Thus, we find that $k = \frac{7!8!}{3!}$, making the desired probability

$$\frac{k}{12!} = \frac{7!8!}{12!3!} = \frac{7}{99}.$$

∎

Second Solution: We assumed in the first solution that each tree is distinguishable. The problem can also be interpreted to mean that trees are distinguishable if and only if they are of different kinds. Under this assumption, the 12 trees can be planted in $\binom{12}{3,4,5} = \frac{12!}{3!4!5!}$ orders.

Let k' be the number of orders in which no two birch trees are adjacent to one another. The probability we need is $\frac{k'}{\binom{12}{3,4,5}}$. To find k', we will count the number of patterns in Figure 2.6, where the seven N's denote non-birch trees, and slots 1 through 8 are occupied by birch trees, with at most one in each slot. We choose three of the seven N's to be maple trees and leave the other four N's for oak trees. There are $\binom{7}{3}$ ways to do this. Then we need to pick five of the eight slots to plant birch trees. There are $\binom{8}{5}$ ways to do so. Hence $k' = \binom{7}{3}\binom{8}{5} = \frac{7!8!}{3!4!5!3!}$, and the desired probability is

$$\frac{k'}{\binom{12}{3,4,5}} = \frac{7!8!}{12!3!} = \frac{7}{99}.$$

∎

Example 2.10. [AIME 2001] Let S be the set of points whose coordinates x, y, and z are integers that satisfy $0 \le x \le 2$, $0 \le y \le 3$, and $0 \le z \le 4$. Two distinct points are randomly chosen from S. What is the probability that the midpoint of the segment they determine also belongs to S?

First Solution: The points of S are lattice points. For a point s in S, there are 3, 4, and 5 possible values for the x-coordinate, y-coordinate, and z-coordinate of s, respectively. Hence $|S| = 3 \cdot 4 \cdot 5 = 60$.

Because $|S| = 60$, there are $\binom{60}{2} = 1770$ ways to choose the two points. In order for the midpoint of the segment joining two chosen points to be a lattice point, it is necessary and sufficient that corresponding coordinates have the same parity. Notice that there are

  $2 \cdot 2 \cdot 3 = 12$ points whose coordinates are all even,
  $1 \cdot 2 \cdot 2 = 4$ points whose coordinates are all odd,
  $1 \cdot 2 \cdot 3 = 6$ points whose only odd coordinate is x,
  $2 \cdot 2 \cdot 3 = 12$ points whose only odd coordinate is y,
  $2 \cdot 2 \cdot 2 = 8$ points whose only odd coordinate is z,
  $2 \cdot 2 \cdot 2 = 8$ points whose only even coordinate is x,
  $1 \cdot 2 \cdot 2 = 4$ points whose only even coordinate is y, and
  $1 \cdot 2 \cdot 3 = 6$ points whose only even coordinate is z.

Thus the desired number of segments is

$$\binom{12}{2} + \binom{4}{2} + \binom{6}{2} + \binom{12}{2} + \binom{8}{2} + \binom{8}{2} + \binom{4}{2} + \binom{6}{2} = 230,$$

and the requested probability is $\dfrac{230}{1770} = \dfrac{23}{177}$. ∎

Second Solution: Because $|S| = 60$, there are $60 \cdot 59 = 3540$ ways to choose the first lattice point and then a distinct second. In order for their midpoint to be a lattice point, it is necessary and sufficient that corresponding coordinates have the same parity. For the x-coordinate, there are 2 ways to be even and one way to be odd. There are $2^2 + 1^2 = 5$ ways for the first coordinates to have the same parity, including 3 ways in which the coordinates are the same. Likewise, there are $2^2 + 2^2 = 8$ ways for the second coordinates to have the same parity, including 4 ways in which the coordinates are the same. There are $3^2 + 2^2 = 13$ ways for the third coordinates to have the same parity, including 5 in which the coordinates are the same. It follows that there are $5 \cdot 8 \cdot 13 - 3 \cdot 4 \cdot 5 = 460$ ways to choose two distinct lattice points, so that the midpoint of the resulting segment is also a lattice point. The requested probability is $\frac{460}{3540} = \frac{23}{177}$. ∎

An interesting and popular incorrect solution goes as follows. We first ignore the condition "distinct." Then there are 60^2 ways to choose two points (with repetition allowed) in order. Then the probability that the midpoints of the two chosen points belong to the set S is $p_1 = \frac{5 \cdot 8 \cdot 13}{60^2} = \frac{13}{90}$. The probability of obtaining an identical pair is $p_2 = \frac{1}{60}$. Hence the answer is $p = p_1 - p_2 = \frac{23}{180}$. Where is the catch? In Figure 2.7, let B be the set of pairs of ordered distinct points in S, and let A be the subset of B containing those pairs of points whose midpoints also in S. We want to calculate $p = \frac{|A|}{|B|}$. Let C be the set of pairs of identical points in S. It is clear that C and B are mutually exclusive and that the midpoints of pairs of points in C belong to S. Then $p_1 = \frac{|A| + |C|}{|B| + |C|}$, $p_2 = \frac{|C|}{|B| + |C|}$, and $p_1 - p_2 = \frac{|A|}{|B| + |C|} \neq p$.

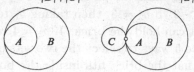

Figure 2.7.

Example 2.11. There are nine pairs of distinct socks in Adrian's drawer. Waking up late for his 9 o'clock class, he randomly grabs eight socks from the drawer without looking. What is the probability that he has exactly two matching pairs of socks among the socks he grabbed?

Solution: There are 18 socks in all. It is clear that there are $\binom{18}{8}$ ways to grab eight socks at random. There are $\binom{9}{2}$ ways to obtain the two matching pairs. Among the seven remaining pairs, four of them have one sock picked. There are $\binom{7}{4}$ ways to pick those pairs. For each of the four pairs, there are two ways to have one sock picked. Hence there are $\binom{9}{2}\binom{7}{4}2^4$ ways to grab eight socks with exactly two matching pairs. It follows that the answer is

$$\frac{\binom{9}{2}\binom{7}{4}2^4}{\binom{18}{8}} = \frac{\frac{9!}{2!\cdot 7!}\frac{7!}{3!\cdot 4!}\cdot 16}{\frac{18!}{8!10!}} = \frac{1120}{2431}.$$

∎

Example 2.12. [AIME 1993] Three numbers a_1, a_2, a_3, are drawn randomly and without replacement from the set $\{1, 2, 3, \ldots, 1000\}$. Three other numbers, b_1, b_2, b_3, are then drawn randomly and without replacement from the remaining set of 997 numbers. Compute the probability that, after a suitable rotation, a brick of dimensions $a_1 \times a_2 \times a_3$ can be enclosed in a box of dimensions $b_1 \times b_2 \times b_3$, with sides of the brick parallel to the sides of the box.

Solution: Since we may rotate the brick before we attempt to place it in the box, we may assume that $a_1 < a_2 < a_3$ and that $b_1 < b_2 < b_3$. The brick will then fit in the box if and only if $a_1 < b_1$, $a_2 < b_2$, and $a_3 < b_3$. Because each selection of six dimensions is equally likely, there is no loss of generality in assuming that the brick and box dimensions are selected from the set $\{1, 2, 3, 4, 5, 6\}$. There are $\binom{6}{3} = 20$ ways to select the dimensions of a brick–box pair from $\{1, 2, 3, 4, 5, 6\}$. If the brick does fit inside the box, then we must have $a_1 = 1$ and $b_3 = 6$. In addition, we must have $b_2 > b_1 > a_1 = 1$ and $b_2 > a_2$, so $6 > b_2 > 3$. If $b_2 = 5$, then taking b_1 to be 2, 3, or 4 will result in a box that can hold the brick. If $b_2 = 4$, we can take $b_1 = 2$ or 3. Thus there are 5 ways to select the two sets of dimensions from $\{1, 2, 3, 4, 5, 6\}$ so that the brick fits inside the box. It follows that the probability that the brick will fit inside the box is $5/20 = 1/4$.

Example 2.13. In how many ways can one arrange 5 indistinguishable armchairs and 5 indistinguishable armless chairs around a circular table (Figure 2.8)? Here two arrangements are considered the same if one can be obtained from the other by rotation.

Solution: If the chairs are arranged in a row, there are $\binom{10}{5} = 252$ ways to arrange them. We number the 10 positions around the table form 1 to 10 in clockwise order. Because this problem involves circular arrangements, we consider position $k + 10$ to be the same as position k; that is, we consider the position of chairs modulo 10.

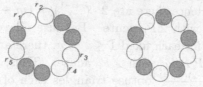

Figure 2.8.

Each arrangement can be rotated, in clockwise direction, by $1, 2, \ldots, 9$ positions. For most arrangements, all of them are different. Hence most arrangements are counted 10 times considering rotations. We need to find those arrangements that repeat themselves after a clockwise rotation of some k positions, where $1 \leq k \leq 9$. Let $r_1, r_2, \ldots, r_5$ be the positions of the armchairs in such an arrangement. Then

$$\{r_1, r_2, r_3, r_4, r_5\} \equiv \{r_1 + k, r_2 + k, r_3 + k, r_4 + k, r_5 + k\} \quad (\text{mod } 10),$$

implying that, modulo 10,

$$r_1 + r_2 + r_3 + r_4 + r_5 \equiv (r_1 + k) + (r_2 + k) + (r_3 + k) + (r_4 + k) + (r_5 + k),$$

or $5k \equiv 0 \pmod{10}$. Consequently, $k = 2, 4, 6, 8$. Now it is not difficult to check that each of these values of k leads to $\{r_1, r_2, r_3, r_4, r_5\} = \{1, 3, 5, 7, 9\}$ or $\{r_1, r_2, r_3, r_4, r_5\} = \{2, 4, 6, 8, 10\}$.

Thus, among the 252 arrangements, two of them are rotations of each other, and each of the other 250 arrangements is counted 10 times considering rotations. Hence the answer is $\frac{250}{10} + \frac{2}{2} = 26$. ∎

Example 2.14. Let n be an integer with $n \geq 3$. Let $P_1 P_2 \ldots P_n$ be a regular n-sided polygon inscribed in a circle ω (Figure 2.9). Three points $P_i, P_j,$ and P_k are randomly chosen, where $i, j,$ and k are distinct integers between 1 and n, inclusive. What is the probability that the triangle $P_i P_j P_k$ is obtuse?

Solution: Because ω is the circumcircle of the triangle $P_i P_j P_k$, the center of ω is outside the triangle if and only if the triangle is obtuse. We consider two cases.

If n is even, say $n = 2m$ for some positive integer $m \geq 2$, then P_{m+1} is diametrically opposite to P_1. We count the number of obtuse triangles $P_1 P_j P_k$ with $\angle P_1$ acute. Then both P_j and P_k must be on the same side of line $P_1 P_{m+1}$. There are $m - 1$ vertices on each side of line $P_1 P_{m+1}$. Hence there are $2 \cdot \binom{m-1}{2} = (m-1)(m-2)$ obtuse triangles $P_1 P_j P_k$ with $\angle P_1$ acute. Because all vertices are evenly distributed on ω, for each fixed $1 \leq i \leq n$, there are $(m-1)(m-2)$ obtuse triangles $P_i P_j P_k$ with $\angle P_i$ acute. Thus, there are $n(m-1)(m-2) = \frac{n(n-2)(n-4)}{4}$ obtuse triangles each of which is counted twice, because there are two acute angles in each obtuse triangle. Consequently, there are $\frac{n(n-2)(n-4)}{8}$ distinct obtuse triangles.

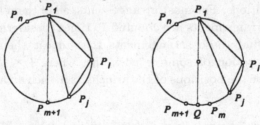

Figure 2.9.

If n is odd, say $n = 2m - 1$ for some positive integer $m \geq 2$, let Q be the point on ω that is diametrically opposite to P_1. Then Q is the midpoint of the minor arc $\overparen{P_m P_{m+1}}$. As in the first case, we can show that there are $2 \cdot \binom{m-1}{2} \cdot n \cdot \frac{1}{2} = \frac{n(n-1)(n-3)}{8}$ distinct obtuse triangles.

There are $\binom{n}{3} = \frac{n(n-1)(n-2)}{6}$ triangles. Thus the desired probability is

$$
\begin{cases}
\dfrac{3(n-4)}{4(n-1)} & \text{if } n \text{ is even,} \\[2mm]
\dfrac{3(n-3)}{4(n-2)} & \text{if } n \text{ is odd.}
\end{cases}
$$

∎

It is interesting to notice that one has a 50% chance of obtaining an obtuse triangle if and only if $n = 5$ or $n = 10$. For large values of n, it is approximately three times as likely to obtain an obtuse triangle than it is to obtain an acute triangle.

Exercises 2

2.1. [AIME 1988] One commercially available ten-button lock may be opened by pressing—in any order—the correct five buttons. Suppose these locks are redesigned so that sets of as many as nine buttons or as few as one button can serve as combinations. How many additional combinations would this allow?

2.2. In the Baseball World Series, the American-League champion and the National-League champion play a best four out of seven series; that is, teams play at most seven games and the series ends when the first team reaches four wins. In 2004, the Exeter Reds will play against the Andover Blues. Suppose they are two evenly matched teams each with a 50% chance of winning any game they play. What is the probability that the Reds will win the series in the seventh game? In how many different win–loss sequences can the Reds win the series?

2.3. Two problems on counting geometric shapes:

  (a) [AHSME 1999] Six points on a circle are given. Four of the chords joining pairs of the six points are selected at random. What is the probability that the four chords form a convex quadrilateral?

  (b) [ARML 2003] A square is tiled by 24 congruent rectangles. Points A and B are the midpoints of a pair of opposite sides of the square. On each side of line AB, four rectangles are chosen at random and colored black. The square is then folded over line AB. Compute the probability that exactly one pair of black rectangles is coincident.

2.4. [AIME 1998] Nine tiles are numbered 1, 2, 3, ... , 9. Each of three players randomly selects and keeps three of the tiles and sums the three values. What is the probability that all three players obtain an odd sum?

2.5. [AIME 1989] When a certain biased coin is flipped 5 times, the probability of getting heads exactly once is not equal to 0 and is the same as that of getting heads exactly twice. What is the probability that the coin comes up heads exactly 3 times out of 5?

2.6. [AIME 2002] Let $A_1, A_2, \ldots, A_{12}$ be the vertices of a regular dodecagon. How many distinct squares in the plane of the dodecagon have at least two vertices in the set $\{A_1, A_2, \ldots, A_{12}\}$?

2.7. Let m, n, k be positive integers with $k \leq m$ and $k \leq n$. Given an $m \times n$ chessboard and k rooks, determine the number of ways to place the rooks on the chessboard such that the rooks do not attack each other.

2.8. One tries to pick three distinct numbers from the set $\{1, 2, \ldots, 34\}$ in such a way that the sum of the three numbers is divisible by three. In how many ways can this be done?

2.9. [AIME 1997] Every card in a deck has a picture in one shape–a circle, a square, or a triangle, which is painted in one of three colors–red, blue, or green. Furthermore, each color is applied in one of three shades–light, medium, or dark. The deck has 27 cards, with every shape–color–shade combination represented. A set of three cards from the deck is called *complementary* if all of the following statements are true:

(a) Either each of the three cards has a different shape or all three cards have the same shape.

(b) Either each of the three cards has a different color or all three cards have the same color.

(c) Either each of the three cards has a different shade or all three cards have the same shade.

How many different complementary three-card sets are there?

2.10. Let m be a positive integer. The numbers $1, 2, \ldots, m$ are evenly spaced around a circle. A red marble is placed next to each number. The marbles are indistinguishable. Adrian wants to choose k marbles ($k \leq \frac{m}{2}$), color them blue, and place them back in their original positions in such a way that there are no neighboring blue marbles in the resulting configuration. In how many ways can he do this?

2.11. Claudia wants to use 8 indistinguishable red beads and 32 indistinguishable blue beads to make a necklace such that there are at least 2 blue beads between any 2 red beads. In how many ways can she do this?

2.12. [IMO 1964] Suppose five points in a plane are situated in such a way that no two of the straight lines joining them are parallel, perpendicular, or coincident. From each point perpendiculars are drawn to all the lines joining the other four points. Determine the maximum number of intersections that these perpendiculars can have.

3
Properties of Binomial Coefficients

Example 3.1. [AIME 1987] A given sequence $r_1, r_2, \ldots, r_n$ of distinct real numbers can be put in ascending order by means of one or more "bubble passes." A bubble pass through a given sequence consists in comparing the second term with the first term and exchanging them if and only if the second term is smaller, then comparing the third term with the current second term and exchanging them if and only if the third term is smaller, and so on, in order, through comparing the last term, r_n, with its current predecessor and exchanging them if and only if the last term is smaller. Figure 3.1 shows how the sequence $1, 9, 8, 7$ is transformed into the sequence $1, 8, 7, 9$ by one bubble pass. The numbers compared at each step are underlined.

$$\underline{1} \quad \underline{9} \quad 8 \quad 7$$
$$1 \quad \underline{9} \quad \underline{8} \quad 7$$
$$1 \quad 8 \quad \underline{9} \quad \underline{7}$$
$$1 \quad 8 \quad 7 \quad 9$$

Figure 3.1.

Suppose that $n = 40$, and that the terms of the initial sequence $r_1, r_2, \ldots, r_{40}$ are distinct from one another and are in random order. Compute the probability that the number that begins as r_{20} will end up, after one bubble pass, in the 30th place (i.e., will have 29 terms on its left and 10 terms on its right).

Solution: The key property of the bubble pass is that immediately after r_k is compared with its predecessor and possibly switched, the current r_k is the largest member of the set $\{r_1, r_2, \ldots, r_k\}$. Also, this

43

set is the same as it was originally, though the order of its elements may be very different. For a number m that is initially before r_{30} in the sequence to end up as r_{30}, two things must happen: m must move into the 30th position when the current r_{29} ($= m$) and r_{30} are compared, and it must not move out of that position when compared to r_{31}. Therefore, by the key property, m must be the largest number in the original $\{r_1, r_2, \ldots, r_{30}\}$, but not larger than the original r_{31}. In other words, of the first 31 numbers originally, the largest must be r_{31} and the second largest must be m, which in our case is r_{20}.

We arrange the numbers $r_1, r_2, \ldots, r_{40}$ in decreasing order to obtain $a_1, a_2, \ldots, a_{40}$. There should be at least one number larger than r_{20} and at least 29 numbers less than r_{20}. Then $a_k = r_{20}$ for some $2 \le k \le 11$. Then $a_1 \le r_{31} \le a_{k-1}$; that is, there are $k - 1$ possible values for r_{31}. All of the $k - 2$ numbers in the set $\{a_1, \ldots, a_{k-1}\} - \{r_{31}\}$ must appear among nine possible numbers $r_{32}, r_{33}, \ldots, r_{40}$ in a certain order. There are $_9P_{(k-2)}$ ways to obtain this result. The remaining $40 - k$ numbers can be arranged in any order. Hence the total number of sequences satisfying the given property is equal to the sum

$$\sum_{k=2}^{11} \left[\binom{k-1}{1} {}_9P_{(k-2)} (40-k)! \right].$$

There are 40! ways to arrange the numbers without restriction. Hence the desired probability is equal to

$$P = \frac{1}{40!} \sum_{k=2}^{11} \left[\binom{k-1}{1} {}_9P_{(k-2)} (40-k)! \right].$$

Is there an easier way of calculating P? The answer is yes. Whatever the first 31 numbers are, there are 31! equally likely ways to order them. Of course, 29! have the largest number in the 31st slot and the second largest in the 20th slot. (The other 29 numbers have 29! equally likely orderings.) Thus the desired probability must be $\frac{29!}{31!} = \frac{1}{930}$. $\blacksquare$

The two methods above reveal the interesting relation:

$$\frac{1}{40!} \sum_{k=2}^{11} \left[\binom{k-1}{1} {}_9P_{(k-2)} (40-k)! \right] = \frac{29!}{31!}. \qquad (*)$$

Of course, this is not a coincidence. In this section, we will discuss more examples of such fundamental properties in counting. We start by introducing the **binomial coefficient**. Let n be a positive integer.

If we expand the two-variable polynomial $(x+y)^n$ as

$$(x+y)^n = a_0 x^n + a_1 x^{n-1} y + a_2 x^{n-2} y^2 + \cdots + a_{n-1} x y^{n-1} + a_n y^n,$$

then for $0 \le k \le n$, a_k is a binomial coefficient.

Example 3.2.   [PEA Math Materials] Bug Fat is at the origin $O = (0,0)$ in the coordinate plane. Fat jumps from lattice point to lattice point, one jump per second, in the plane following the pattern: From (m,n), he jumps to either $(m, n+1)$ or $(m+1, n)$, each equally likely. Where can Fat be five seconds after he leaves the origin? Are all of these places equally likely?

Solution: We call a jump from (m,n) to $(m+1, n)$ an r jump (a right jump) and a jump from (m,n) to $(m, n+1)$ a u jump (an upper jump). It is clear that Fat can reach $(5,0)$, $(4,1)$, $(3,2)$, $(2,3)$, $(1,4)$, or $(0,5)$ in five seconds, since for nonnegative integers m and n, it takes $m+n$ seconds to reach (m,n). Let $P_{(m,n)}$ denote the number of ways by which Fat can reach (m,n) in $m+n$ seconds. To reach the point (m,n), Fat needs to reach either $(m-1, n)$ or $(m, n-1)$. For each way by which Fat reaches $(m-1, n)$ or $(m, n-1)$, he can then make a jump to reach (m,n). Hence $P_{(m,n)} = P_{(m-1,n)} + P_{(m,n-1)}$ (unless m or n is 0). We have the situation of Figure 3.2.

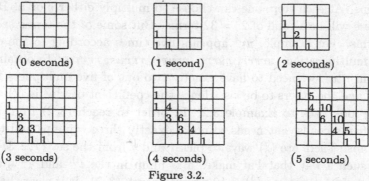

(0 seconds)   (1 second)   (2 seconds)

(3 seconds)   (4 seconds)   (5 seconds)

Figure 3.2.

Therefore, Fat can reach $(5,0)$, $(4,1)$, $(3,2)$, $(2,3)$, $(1,4)$, and $(0,5)$ along 1, 5, 10, 10, 5, and 1 different paths, respectively. There is a total of 32 paths. (Is it a coincidence that $32 = 2^5$?) It is not equally likely for Fat to reach each of these points. The probabilities of reaching $(5,0)$, $(4,1)$, $(3,2)$, $(2,3)$, $(1,4)$, and $(0,5)$ are $\frac{1}{32}$, $\frac{5}{32}$, $\frac{10}{32} = \frac{5}{16}$, $\frac{10}{32} = \frac{5}{16}$, $\frac{5}{32}$, and $\frac{1}{32}$, respectively.   ∎

Putting everything together, we obtain a triangular table of numbers (Figure 3.3). Note that each row starts and ends with 1's and every number in the interior of the triangle is the sum of the two numbers directly above it. Continuing in this pattern, we obtain the famous **Pascal's triangle**.

$$
\begin{array}{ccccccccccccc}
 & & & & & & 1 & & & & & & \\
 & & & & & 1 & & 1 & & & & & \\
 & & & & 1 & & 2 & & 1 & & & & \\
 & & & 1 & & 3 & & 3 & & 1 & & & \\
 & & 1 & & 4 & & 6 & & 4 & & 1 & & \\
 & 1 & & 5 & & 10 & & 10 & & 5 & & 1 & \\
1 & & 6 & & 15 & & 20 & & 15 & & 6 & & 1 \\
 & & \cdots & & & & \cdots & & & & \cdots & &
\end{array}
$$

Figure 3.3.

By convention, the top row of the triangle is considered to be the 0th row.

Now, expanding $(r + u)^5$ yields

$$r^5 + 5r^4u + 10r^3u^2 + 10r^2u^3 + 5ru^4 + u^5.$$

We notice that the coefficients $1, 5, 10, 10, 5, 1$ are identical to those obtained in Example 3.2. Indeed, to expand $(r + u)^5$, one needs to multiply $r + u$ five times. We view these five multiplications as five steps. At each step, one can choose to multiply either r or u. Hence there will be a total of $2^5 = 32$ terms. But some of the terms are like terms. For example, ur^4 appears five times according to the order of multiplication: $urrrr, rurrr, rrurr, rrrur, rrrru$. To obtain the term r^3u^2, we need to have exactly two out of five multipliers to be u's and the others to be r's. Hence the coefficient of r^3u^2 is $\binom{5}{2} = 10$. We look back to Example 3.2. In order to reach $(3, 2)$ from $(0, 0)$ in five seconds, Fat needs to make exactly three r jumps and two u jumps. There are $\binom{5}{2}$ ways to pick i and j from the set $\{1, 2, 3, 4, 5\}$, in such a way that Fat makes a u jump in the ith and j^{th} jumps. Therefore, there are $\binom{5}{2} = 10$ ways to reach $(3, 2)$. It is clear that the actual numbers 5 and 2 are not significant at all. In exactly the same way, we can prove the following theorem.

Theorem 3.1. *Let n be a positive integer. Then*

$$(x + y)^n = \binom{n}{0}x^n + \binom{n}{1}x^{n-1}y + \binom{n}{2}x^{n-2}y^2 + \cdots + \binom{n}{n}y^n.$$

By convention, we set $\binom{n}{0} = \binom{n}{n} = \frac{n!}{n!0!} = 1$. Binomial coefficients are also entries in Pascal's triangle. More precisely, for $n \geq 0$,

$$\binom{n}{0}, \ \binom{n}{1}, \ \binom{n}{2}, \ \ldots, \ \binom{n}{n-1}, \ \binom{n}{n}$$

is the nth row of the triangle.

This theorem explains why the numbers $\binom{n}{k}$ are called **binomial numbers** or **binomial coefficients**. The idea of expanding polynomials and using their coefficients to count objects is very important in combinatorics. Such polynomials are called **generating functions**. We will discuss such functions more carefully in Chapter 8.

Example 3.3. [AIME 1995] Starting at $(0,0)$, an object moves in the coordinate plane via a sequence of steps, each of length one. Each step is left, right, up, or down, all four being equally likely. What is the probability that the object reaches $(2,2)$ in six or fewer steps?

Solution: Since the net movement must be two steps right (R) and two steps up (U), there must be at least four steps. The point $(2,2)$ can be reached in exactly four steps if the sequence is some permutation of R, R, U, U. These four steps can be permuted in $\frac{4!}{2!2!} = 6$ ways. Each of these sequences has probability $(1/4)^4$ of occurring. Thus the probability of reaching $(2,2)$ in exactly 4 steps is $6/4^4$.

In moving to $(2,2)$, the total number of steps must be even, since an odd number of steps would reach a lattice point with one even coordinate and one odd coordinate. Next consider the possibility of reaching $(2,2)$ in six steps. A six-step sequence must include the steps R, R, U, U in some order, as well as a pair consisting of R, L (left) or U, D (down), in some order. The steps R, R, U, U, U, D can be permuted in $\frac{6!}{3!2!1!} = 60$ ways, but for 12 of these sequences, namely, those that start with some permutation of R, R, U, U, the object actually reaches $(2,2)$ in four steps. A similar analysis holds for the steps R, R, U, U, R, L. Thus there are $2(60-12) = 96$ sequences that reach $(2,2)$ but that do not do so until the sixth step. Each of these 96 sequences occurs with probability $1/4^6$.

Considering the four- and six-step possibilities, we find that the probability of reaching $(2,2)$ in six or fewer steps is

$$\frac{6}{4^4} + \frac{96}{4^6} = \frac{3}{64}.$$

Example 3.4. A license plate consists of 8 digits. It is called *even* if it contains an even number of 0's. Find the number of even license plates.

Solution: For $0 \le k \le 4$, if there are $2k$ 0's in the plate, then there are $8 - 2k$ nonzero digits, each of which has 9 choices. There are $\binom{8}{2k}$ ways to choose $2k$ positions for 0's, and $\binom{8}{2k}9^{8-2k}$ plates have exactly $2k$ 0's. Hence the answer is $9^8 + \binom{8}{2}9^6 + \binom{8}{4}9^4 + \binom{8}{6}9^2 + \binom{8}{8}$. By Theorem 3.1, $(9+1)^8 = 9^8 + \binom{8}{1}9^7 + \binom{8}{2}9^6 + \binom{8}{3}9^5 + \binom{8}{4}9^4 + \binom{8}{5}9^3 + \binom{8}{6}9^2 + \binom{8}{7}9 + 1$ and $(9-1)^8 = 9^8 - \binom{8}{1}9^7 + \binom{8}{2}9^6 - \binom{8}{3}9^5 + \binom{8}{4}9^4 - \binom{8}{5}9^3 + \binom{8}{6}9^2 - \binom{8}{7}9 + 1$. Therefore, the answer is $\frac{(9+1)^8 + (9-1)^8}{2} = \frac{10^8 + 8^8}{2}$. ∎

This problem can be also solved using recursive relations. Similar problems will be discussed in Chapter 6.

Theorem 3.2. *Let n and k be positive integers with $n \ge k$. The following fundamental properties of the binomial coefficients $\binom{n}{k}$ hold:*

(a) $\binom{n}{k} = \binom{n}{n-k}$;

(b) $\binom{n}{k+1} = \binom{n-1}{k+1} + \binom{n-1}{k}$;

(c) $\binom{n}{0} < \binom{n}{1} < \binom{n}{2} < \cdots < \binom{n}{\lceil \frac{n-1}{2} \rceil} = \binom{n}{\lfloor \frac{n}{2} \rfloor}$;

(d) $k\binom{n}{k} = n\binom{n-1}{k-1}$;

(e) $k\binom{n}{k} = (n - k + 1)\binom{n}{k-1}$;

(f) $\binom{n}{0} + \binom{n+1}{1} + \binom{n+2}{2} + \cdots + \binom{n+k}{k} = \binom{n+k+1}{k}$;

(g) $\binom{n}{n} + \binom{n+1}{n} + \binom{n+2}{n} + \cdots + \binom{n+k}{n} = \binom{n+k+1}{n+1}$;

(h) $\binom{n}{0} + \binom{n}{1} + \cdots + \binom{n}{n} = 2^n$;

(i) $\binom{n}{0} - \binom{n}{1} + \binom{n}{2} - \cdots + (-1)^n \binom{n}{n} = 0$;

(j) $\binom{n}{1} + 2\binom{n}{2} + 3\binom{n}{3} + \cdots + n\binom{n}{n} = n2^{n-1}$;

(k) $\binom{n}{k}$ is divisible by n if n is prime and $1 \le k \le n - 1$.

Proof: Properties (a) and (b) can be easily derived using the relation $\binom{n}{k} = \frac{n!}{k!(n-k)!}$. We instead present combinatorial arguments to prove them.

(a) One notices that picking a combination of k objects to be kept is equivalent to picking a combination of $n - k$ objects to be left out, implying that $\binom{n}{k} = \binom{n}{n-k}$. This property also follows from the symmetry of the expansion of $(x + y)^n$ with respect to x and y.

(b) This follows directly from the definition of Pascal's triangle. One also notices that the left-hand side denotes the number of ways in which we are able to pick combinations of $k+1$ objects from n given objects. One can also count them in the following way. Let us give one particular object a special name, say Fat. We classify all the $(k+1)$-object combinations into two types. Type (A) consists of those containing Fat. Type (B) consists of those not containing Fat. Then it is not difficult to see that there are $\binom{n-1}{k}$ combinations of type (A) and $\binom{n-1}{k+1}$ combinations of type (B). Since the two results must be the same, $\binom{n}{k+1} = \binom{n-1}{k+1} + \binom{n-1}{k}$.

(c) If n is even, $\left\lceil \frac{n-1}{2} \right\rceil = \frac{n}{2}$. If n is odd, $\left\lceil \frac{n-1}{2} \right\rceil = \frac{n-1}{2}$. For $0 \le k \le \left\lceil \frac{n-1}{2} \right\rceil - 1$, $2k \le n-2$, or $k+1 \le n-k-1$. Hence

$$\frac{\binom{n}{k}}{\binom{n}{k+1}} = \frac{(k+1)!(n-k-1)!}{k!(n-k)!} = \frac{k+1}{n-k} < 1.$$

By (a), we have

$$\binom{n}{0} < \binom{n}{1} < \cdots < \binom{n}{m} = \binom{n}{m+1} > \binom{n}{m+2} > \cdots > \binom{n}{n}$$

if n is odd and $n = 2m+1$, and

$$\binom{n}{0} < \binom{n}{1} < \cdots < \binom{n}{m} > \binom{n}{m+1} > \cdots > \binom{n}{n}$$

if n is even and $n = 2m$.

(d)

$$k\binom{n}{k} = \frac{k \cdot n!}{k!(n-k)!} = \frac{n \cdot (n-1)!}{(k-1)![(n-1)-(k-1)]!} = n\binom{n-1}{k-1}.$$

(e)

$$k\binom{n}{k} = \frac{k \cdot n!}{k!(n-k)!} = \frac{(n-k+1) \cdot n!}{(k-1)!(n-k+1)!}$$

$$= (n-k+1)\binom{n}{k-1}.$$

(f) Note that $\binom{n}{0} = \binom{n+1}{0}$. The desired result follows by using (b) repeatedly. (Indeed, this is what happened in Example 3.1.)

(g) This follows from using (a) on each term of (f).

(h) Setting $x = y = 1$ in the expansion of $(x+y)^n$ gives the desired result. This is also Theorem 1.3. For $0 \le k \le n$, there are $\binom{n}{k}$

k-element subsets of $S = \{1, 2, \ldots, n\}$. Summing up from $k = 0$ to $k = n$ gives the total number of subsets of S (including $\emptyset$ and S itself).

(i) Setting $x = 1$ and $y = -1$ in the expansion of $(x + y)^n$ gives the desired result.

(j) By (d) and (h), we have

$$\sum_{k=1}^{n} k \binom{n}{k} = n \sum_{k=1}^{n} \binom{n-1}{k-1} = n \sum_{k=0}^{n-1} \binom{n-1}{k} = n2^{n-1},$$

as desired.

(k) Note that $\binom{n}{k}$ is an integer. Note also that n divides $n!$, the numerator of $\binom{n}{k}$. If n is prime, n is relatively prime to $k!(n-k)!$. Hence $\binom{n}{k}$ is divisible by n.

The binomial coefficient $\binom{n}{k}$ has a combinatorial meaning if n and k are integers with $0 < k \le n$. In Theorem 3.1, we extend the definition to $k = 0$. We also set $\binom{0}{0} = 1$. If either $0 \le n < k$ or $k < 0 \le n$, one cannot pick k objects out of n objects, so we set $\binom{n}{k} = 0$. It is not difficult to check that all the properties in Theorems 3.1 and 3.2 are satisfied with this generalized definition of the binomial coefficients. Thus $\binom{n}{k}$ is now well defined for integers n and k with $n \ge 0$. From now on, we use the general definition.

Example 3.5. [AIME 1989] Ten points are marked on a circle. How many distinct convex polygons of three or more sides can be drawn using some (or all) of the ten points as vertices? (Polygons are distinct unless they have exactly the same vertices.)

Solution: For $3 \le k \le 10$, each choice of k points will yield a convex polygon with k vertices. Because k points can be chosen from 10 in $\binom{10}{k}$ ways, the answer to the problem is

$$\binom{10}{3} + \binom{10}{4} + \cdots + \binom{10}{10}$$

$$= \left[\binom{10}{0} + \binom{10}{1} + \cdots + \binom{10}{10}\right] - \left[\binom{10}{0} + \binom{10}{1} + \binom{10}{2}\right]$$

$$= (1+1)^{10} - (1 + 10 + 45) = 1024 - 56 = 968,$$

by Theorem 3.2 (h). A careful reader might want to ask: Where have we used the stipulation that the polygons are convex? ∎

Example 3.6. Evaluate

$$\frac{\binom{11}{0}}{1} + \frac{\binom{11}{1}}{2} + \frac{\binom{11}{2}}{3} + \cdots + \frac{\binom{11}{11}}{12}.$$

Solution: By Theorem 3.2 (d), for $0 \le k \le 11$, $\frac{12}{k+1}\binom{11}{k} = \binom{12}{k+1}$, or

$$\frac{\binom{11}{k}}{k+1} = \frac{\binom{12}{k+1}}{12}.$$

Hence the desired sum is equal to

$$\frac{1}{12} \sum_{k=0}^{11} \binom{12}{k+1} = \frac{1}{12} \left(\sum_{k=0}^{12} \binom{12}{k} - \binom{12}{0} \right) = \frac{1}{12}(2^{12} - 1),$$

by Theorem 3.2 (h). ∎

Example 3.7. Let n be a positive integer. Prove that

$$\sum_{k=1}^{n} \frac{(-1)^{k-1}}{k} \binom{n}{k} = 1 + \frac{1}{2} + \cdots + \frac{1}{n}.$$

Solution: Let

$$S_n = \sum_{k=1}^{n} \frac{(-1)^{k-1}}{k} \binom{n}{k}.$$

We have

$$S_{n+1} = \sum_{k=1}^{n+1} \frac{(-1)^{k-1}}{k} \binom{n+1}{k} = \sum_{k=1}^{n+1} \frac{(-1)^{k-1}}{k} \left[\binom{n}{k} + \binom{n}{k-1} \right]$$

$$= \sum_{k=1}^{n} \frac{(-1)^{k-1}}{k} \binom{n}{k} + \sum_{k=1}^{n+1} \frac{(-1)^{k-1}}{k} \binom{n}{k-1}$$

$$= S_n + \sum_{k=1}^{n+1} \frac{(-1)^{k-1}}{n+1} \binom{n+1}{k}$$

$$= S_n - \frac{1}{n+1} \sum_{k=1}^{n+1} (-1)^k \binom{n+1}{k}$$

by Theorem 3.2 (b) and (d). Hence for all $m \geq 1$,

$$S_{m+1} - S_m = -\frac{1}{m+1} \sum_{k=1}^{m+1} (-1)^k \binom{m+1}{k}$$

$$= \frac{1}{m+1} - \frac{1}{m+1} \sum_{k=0}^{m+1} (-1)^k \binom{m+1}{k} = \frac{1}{m+1},$$

by Theorem 3.2 (i). Summing from $m = 1$ to $m = n - 1$ yields

$$S_n - S_1 = \frac{1}{2} + \cdots + \frac{1}{n},$$

from which the desired result follows by taking into account that $S_1 = 1$. ∎

Now we revisit $(*)$ in Example 3.1. Note that

$$\sum_{k=2}^{11} \left[\binom{k-1}{1} {}_9P_{(k-2)}(40-k)! \right] = \sum_{k=2}^{11} \frac{(k-1) \cdot 9!(40-k)!}{(11-k)!}$$

$$= \sum_{k=0}^{9} \frac{(k+1) \cdot 9!(38-k)!}{(9-k)!}.$$

Hence $(*)$ is equivalent to

$$\sum_{k=0}^{9} \frac{(k+1) \cdot (38-k)!}{29!(9-k)!} = \frac{40!}{31!9!},$$

or

$$\sum_{k=0}^{9} (k+1) \binom{38-k}{9-k} = \binom{40}{9}.$$

Note that

$$\sum_{k=0}^{9} (k+1) \binom{38-k}{9-k} = \sum_{k=0}^{9} (10-k) \binom{29+k}{k}$$

$$= 10 \sum_{k=0}^{9} \binom{29+k}{k} - \sum_{k=0}^{9} k \binom{29+k}{k}.$$

By Theorem 3.2 (f),

$$\sum_{k=0}^{9} \binom{29+k}{k} = \binom{39}{9}.$$

By Theorem 3.2 (e), $k\binom{29+k}{k} = 30\binom{29+k}{k-1}$, so

$$\sum_{k=0}^{9} k\binom{29+k}{k} = 30\sum_{k=0}^{9}\binom{29+k}{k-1} = 30\binom{39}{8}$$

by (f). Now it suffices to show that

$$10\binom{39}{9} - 30\binom{39}{8} = \binom{40}{9}.$$

By (e) and then by (a), we have

$$10\binom{39}{9} - 30\binom{39}{8} = \binom{39}{9} + 9\binom{39}{9} - 30\binom{39}{8}$$

$$= \binom{39}{9} + 31\binom{39}{8} - 30\binom{39}{8}$$

$$= \binom{39}{9} + \binom{39}{8} = \binom{40}{9},$$

as desired. Readers are encouraged to find and prove the general form of $(*)$.

Example 3.8.   Let $\{F_n\}_{n=0}^{\infty}$ be the sequence defined by $F_0 = F_1 = 1$ and $F_{n+2} = F_{n+1} + F_n$ for $n \geq 0$. Prove that

$$\sum_{k=0}^{n}\binom{n-k+1}{k} = F_{n+1}.$$

The sequence $\{F_n\}_{n=0}^{\infty}$ is called the **Fibonacci sequence** and F_n the nth **Fibonacci number**.

Solution: For $n \geq 0$, let

$$f_n = \sum_{k=0}^{n}\binom{n-k+1}{k}.$$

It is easy to check that $f_0 = \binom{1}{0} = 1 = F_1$ and $f_1 = \binom{2}{0} + \binom{1}{1} = 2 = F_2$. It suffices to show that $f_{n+2} = f_{n+1} + f_n$ for $n \geq 0$, because the sequence will then be completely determined by the recursive relation and its initial values f_1 and f_2. By Theorem 3.2 (b), we have

$$f_{n+2} = \sum_{k=0}^{n+2} \binom{n-k+3}{k} = \sum_{k=0}^{n+2} \left[\binom{n-k+2}{k} + \binom{n-k+2}{k-1} \right]$$

$$= \sum_{k=0}^{n+2} \binom{n-k+2}{k} + \sum_{k=0}^{n+2} \binom{n-k+2}{k-1}$$

$$= \sum_{k=0}^{n+1} \binom{n-k+2}{k} + \binom{0}{n+2}$$

$$\quad + \sum_{k=1}^{n+1} \binom{n-k+2}{k-1} + \binom{n+2}{-1} + \binom{0}{n+1}$$

$$= \sum_{k=0}^{n+1} \binom{(n+1)-k+1}{k} + 0 + \sum_{k=0}^{n} \binom{n-k+1}{k} + 0 + 0$$

$$= f_{n+1} + f_n,$$

as desired. ∎

Figure 3.4 illustrates the relation between f_4, f_5, and f_6, where Theorem 3.2 (b) has been used four times. Indeed, $f_4 = 1 + 4 + 3$, $f_5 = 1+5+6+1$, and $f_6 = 1+6+10+4 = 1+(1+5)+(4+6)+(3+1) = f_4 + f_5$.

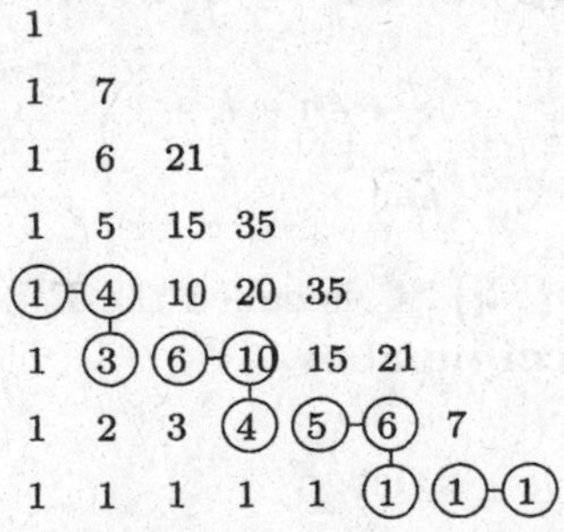

$$
\begin{array}{ccccccccc}
1 \\
1 & 7 \\
1 & 6 & 21 \\
1 & 5 & 15 & 35 \\
① & ④ & 10 & 20 & 35 \\
1 & ③ & ⑥ & ⑩ & 15 & 21 \\
1 & 2 & 3 & ④ & ⑤ & ⑥ & 7 \\
1 & 1 & 1 & 1 & 1 & ① & ① & ① \\
\end{array}
$$

Figure 3.4.

Example 3.9. [TST 2000, from Kvant] Let n be a positive integer. Prove that

$$\sum_{i=0}^{n} \binom{n}{i}^{-1} = \frac{n+1}{2^{n+1}} \sum_{i=1}^{n+1} \frac{2^i}{i}.$$

Solution: Let

$$S_n = \frac{1}{n+1} \sum_{i=0}^{n} \binom{n}{i}^{-1} = \frac{1}{n+1} \sum_{i=0}^{n} \frac{i!(n-i)!}{n!}.$$

We must show that $2^{n+1} S_n = \sum_{i=1}^{n+1} \frac{2^i}{i}$. To do so, it suffices to check that $S_1 = 1$, which is clear, and that $2^{n+2} S_{n+1} - 2^{n+1} S_n = \sum_{i=1}^{n+2} \frac{2^i}{i} - \sum_{i=1}^{n+1} \frac{2^i}{i} = \frac{2^{n+2}}{n+2}$, or $2S_{n+1} = S_n + \frac{2}{n+2}$. Now,

$$2S_{n+1} = \frac{1}{n+2} \left(\sum_{i=0}^{n+1} \binom{n+1}{i}^{-1} + \sum_{i=0}^{n+1} \binom{n+1}{i}^{-1} \right)$$

$$= \frac{2}{n+2} + \frac{1}{n+2} \sum_{i=0}^{n} \left(\binom{n+1}{i}^{-1} + \binom{n+1}{i+1}^{-1} \right)$$

$$= \frac{2}{n+2} + \frac{1}{n+2} \sum_{i=0}^{n} \frac{i!(n+1-i)! + (i+1)!(n-i)!}{(n+1)!}$$

$$= \frac{2}{n+2} + \frac{1}{n+2} \sum_{i=0}^{n} \frac{i!(n-i)!(n+1-i+i+1)}{(n+1)!}$$

$$= \frac{2}{n+2} + \frac{1}{n+1} \sum_{i=0}^{n} \frac{i!(n-i)!}{n!}$$

$$= S_n + \frac{2}{n+2},$$

as claimed.   ∎

In this problem we found a *reciprocal* version of Theorem 3.2 (b); that is,

$$\binom{n+1}{k}^{-1} + \binom{n+1}{k+1}^{-1} = \frac{n+2}{n+1} \binom{n}{k}^{-1}.$$

Example 3.10. [Vandermonde] Let m, n, and k be integers with $m, n \geq 0$. Prove that

$$\binom{m+n}{k} = \sum_{i=0}^{k} \binom{m}{i} \binom{n}{k-i}.$$

Solution: Assume that there are m male dorms and n female dorms at PEA. We want to pick k dorms at random to check the dorms' fire hazards. Of course, there are $\binom{m+n}{k}$ ways to do so. On the other hand, these picks can be classified into $k + 1$ types: For $0 \leq i \leq k$, the ith type are those consisting of i male dorms and $k - i$ female dorms. Thus there are $\binom{m}{i}\binom{n}{k-i}$ picks of the ith type. Summing these numbers from 0 to k gives the desired identity.   ■

The above approach is one of the most intriguing methods in combinatorics: the combinatorial model. Counting the model in two ways yields the desired identity. In Chapter 8, we will present another method, which uses generating functions. Setting special values for m, n, and k in the Vandermonde identity leads to many interesting results. For example, $m = n = k$ gives

$$\binom{n}{0}^2 + \binom{n}{1}^2 + \cdots + \binom{n}{n}^2 = \binom{2n}{n}.$$

In some situations, a very effective approach is to view $\binom{n}{k}$ as a function of k. Not only does this simplify many recursive calculations, but it also allows us to apply the theory of polynomials and functions.

Example 3.11. Let n and k be positive integers with $k + 3 \leq n$. Prove that $\binom{n}{k}$, $\binom{n}{k+1}$, $\binom{n}{k+2}$, $\binom{n}{k+3}$ cannot form an arithmetic progression.

Solution: Assume to the contrary that there are positive integers n and k such that $a_1 = \binom{n}{k}$, $a_2 = \binom{n}{k+1}$, $a_3 = \binom{n}{k+2}$, $a_4 = \binom{n}{k+3}$ form an arithmetic progression. Setting

$$f(k) = \binom{n}{k} + \binom{n}{k+2} - 2\binom{n}{k+1}$$

gives $f(k) = f(k+1) = 0$, because $a_1 + a_3 - 2a_2 = a_2 + a_4 - 2a_3 = 0$. Note that

$$f(k) = \frac{n!}{k!(n-k)!} + \frac{n!}{(k+2)!(n-k-2)!} - \frac{2 \cdot n!}{(k+1)!(n-k-1)!}.$$

Hence $f(k) = 0$ implies

$$\frac{n!}{(k+2)!(n-k)!}[(k+1)(k+2) + (n-k-1)(n-k) - 2(n-k)(k+2)] = 0,$$

and thus

$$g(k) = (k+1)(k+2) + (n-k-1)(n-k) - 2(n-k)(k+2) = 0.$$

Consider the quadratic equation

$$g(x) = (x+1)(x+2) + (n-x-1)(n-x) - 2(n-x)(x+2) = 0.$$

Then $g(x) = 0$ can have at most two roots. Since $f(k) = f(k+1) = 0$ implies that $g(k) = g(k+1) = 0$ we conclude that $x = k$ and $x = k+1$ are the two roots of $g(x) = 0$.

By Theorem 3.2 (a),

$$f(n-k-2) = \binom{n}{n-k-2} + \binom{n}{n-k} - 2\binom{n}{n-k-1}$$
$$= \binom{n}{k} + \binom{n}{k+2} - 2\binom{n}{k+1} = f(k).$$

Hence $g(n-k-2) = 0$. Likewise, $g(n-(k+1)-2) = g(n-k-3) = 0$. Therefore, $k, k+1, n-k-3$, and $n-k-2$ are roots of the quadratic equation $g(x) = 0$. Hence $k = n-k-3$ and $n = 2k+3$. Then a_1, a_2, a_3, a_4 are the four middle terms of the nth row of Pascal's triangle. They cannot form an arithmetic progression by Theorem 3.2 (c). Thus our original assumption was wrong, and the problem statement is true.

Example 3.12. [AIME Proposal] In a round robin tournament, a cyclic 3-set occurs occasionally; that is, a set $\{a, b, c\}$ of three teams where a beats b, b beats c, and c beats a. If 23 teams play a round robin tournament (without ties), what is the largest number of cyclic 3-sets that can occur?

Solution: We first consider the number of 3-sets that are not cyclic. Note that a set $\{a, b, c\}$ of three teams is not cyclic if and only if there is a team that beats the other two teams. Note also that a set $\{a, b, c\}$ of three teams is not cyclic if and only if there is a team that loses to the other two teams. Let $t_1, t_2, \ldots, t_{23}$ denote all the teams, and assume that team t_i beats w_i teams and loses to v_i teams. Hence the number of non-cyclic 3-sets is

$$M = \sum_{i=1}^{23} \binom{w_i}{2} = \sum_{i=1}^{23} \binom{v_i}{2} = \frac{1}{2} \sum_{i=1}^{23} \left(\binom{w_i}{2} + \binom{v_i}{2} \right).$$

For each i,

$$\binom{w_i}{2} + \binom{v_i}{2} = \frac{w_i(w_i - 1)}{2} + \frac{v_i(v_i - 1)}{2}$$

$$= \frac{1}{2}[(w_i^2 + v_i^2) - (w_i + v_i)]$$

$$\geq \frac{1}{2}\left[\frac{(w_i + v_i)^2}{2} - (w_i + v_i)\right]$$

by the **(Quadratic) Root Mean Square - Arithmetic Mean Inequality**. Since $w_i + v_i = 22$ for each i,

$$\binom{w_i}{2} + \binom{v_i}{2} \geq \frac{1}{2}[242 - 22] = 110,$$

and there are at least $\frac{1}{2} \cdot 23 \cdot 110 = 1265$ noncyclic 3-sets. Consequently, there are at most

$$\binom{23}{3} - 1265 = 506$$

cyclic 3-sets. Equality holds if and only if $w_i = v_i = 11$ for each i. It remains to be shown that such an outcome is possible. To see this, we arrange all 23 teams around a round table and fix the matches so that each team beats the first 11 teams sitting clockwise from it. ∎

In this example, we used the **convexity** of $\binom{n}{k}$ as a function in k. More precisely, let m denote the minimum value of M for

(i) $0 \leq w_i \leq 22$;

(ii) $\sum_{i=1}^{23} w_i = \binom{23}{2}$.

We claim that m is achieved if and only if $w_i = \frac{1}{23}\binom{23}{2} = 11$. Assume to the contrary that some $w_k > 11$. Then there is a j such that $w_j < 11$. Change w_k to $w_k - 1$ and w_j to $w_j + 1$, leaving the other w_i's intact. Note that the new sequence still satisfies conditions (i) and (ii), while it reduces the value of $\sum_{i=1}^{23} \binom{w_i}{2}$ because

$$\binom{w_k - 1}{2} + \binom{w_j + 1}{2} = \binom{w_k}{2} - (w_k - 1) + \binom{w_j}{2} + w_j < \binom{w_k}{2} + \binom{w_j}{2},$$

since $w_j + 1 \leq 11 < w_k$. It follows that this new sequence leads to a smaller value than the assumed minimal value. Hence our assumption was wrong, and for sequences satisfying conditions (i) and (ii), $M = m$ if and only if $w_i = 11$ for each i.

The following theorems by E. Lucas in 1878 and E. Kummer in 1852 are very useful in number theory. Let n be a positive integer, and let p be a prime. Let $(\overline{n_m n_{m-1} \ldots n_0})_p$ denote the **base p representation of n**; that is,

$$n = (\overline{n_m n_{m-1} \ldots n_0})_p = n_0 + n_1 p + \cdots + n_m p^m,$$

where $0 \le n_0, n_1, \ldots, n_m \le p - 1$ and $n_m \ne 0$.

Theorem 3.3. *[Lucas] Let p be a prime, and let n be a positive integer with $n = (\overline{n_m n_{m-1} \ldots n_0})_p$. Let i be a positive integer less than n. Also, write $i = i_0 + i_1 p + \cdots + i_m p^m$, where $0 \le i_0, i_1, \ldots, i_m \le p - 1$. Then*

$$\binom{n}{i} \equiv \prod_{j=0}^{m} \binom{n_j}{i_j} \quad (\bmod\ p). \tag{$**$}$$

Here $\binom{0}{0} = 1$ and $\binom{n_j}{i_j} = 0$ if $n_j < i_j$ by the convention we set earlier.

To prove this theorem, we need some new terminology. Let p be a prime, and let $f(x)$ and $g(x)$ be two polynomials with integer coefficients. We say that $f(x)$ is **congruent** to $g(x)$ modulo p, and write $f(x) \equiv g(x) \pmod{p}$ if all of the coefficients of $f(x) - g(x)$ are divisible by p. (Note that the congruence of polynomials is different from the congruence of the values of polynomials. For example, $x(x + 1) \not\equiv 0 \pmod 2$ even though $x(x + 1)$ is divisible by 2 for all integers x.) The following properties can easily be verified:

(a) $f(x) \equiv f(x) \pmod{p}$;

(b) if $f(x) \equiv g(x) \pmod{p}$, then $g(x) \equiv f(x) \pmod{p}$;

(c) if $f(x) \equiv g(x) \pmod{p}$ and $g(x) \equiv h(x) \pmod{p}$, then $f(x) \equiv h(x) \pmod{p}$;

(d) if $f(x) \equiv g(x) \pmod{p}$ and $f_1(x) \equiv g_1(x) \pmod{p}$, then $f(x) \pm f_1(x) \equiv g(x) \pm g_1(x) \pmod{p}$ and $f(x)f_1(x) \equiv g(x)g_1(x) \pmod{p}$.

In other words, the congruence relation between polynomials is reflexive, symmetric, and transitive (so it is an equivalence relation), and is preserved under addition, subtraction, and multiplication.

Proof: By Theorem 3.2 (k), the binomial coefficients $\binom{p}{k}$, where $1 \le k \le p - 1$, is divisible by p. Thus, $(1 + x)^p \equiv 1 + x^p \pmod{p}$ and $(1 + x)^{p^2} = [(1 + x)^p]^p \equiv [1 + x^p]^p \equiv 1 + x^{p^2} \pmod{p}$, and so on; so that for any positive integer r, $(1 + x)^{p^r} \equiv 1 + x^{p^r} \pmod{p}$ by induction. We consider $(1 + x)^n$.

We have

$$(1 + x)^n = (1 + x)^{n_0 + n_1 p + \cdots + n_m p^m}$$

$$= (1 + x)^{n_0} [(1 + x)^p]^{n_1} \cdots [(1 + x)^{p^m}]^{n_m}$$

$$\equiv (1 + x)^{n_0} (1 + x^p)^{n_1} \cdots (1 + x^{p^m})^{n_m} \pmod{p}.$$

The coefficient of x^i in the expansion of $(1+x)^n$ is $\binom{n}{i}$. On the other hand, because $i = i_0 + i_1 p + \cdots + i_m p^m$, the coefficient of x^i is the coefficient of $x^{i_0} (x^p)^{i_1} \cdots (x^{p^m})^{i_m}$, which is equal to $\binom{n_0}{i_0} \binom{n_1}{i_1} \cdots \binom{n_m}{i_m}$. Hence

$$\binom{n}{i} \equiv \binom{n_0}{i_0} \binom{n_1}{i_1} \cdots \binom{n_m}{i_m} \pmod{p},$$

as desired.   ∎

The proof of this theorem is yet another good application of generating functions, which will be discussed extensively in chapter 8.

Theorem 3.4. *[Kummer] Let n and i be positive integers with $i \leq n$, and let p be a prime. Then p^t divides $\binom{n}{i}$ if and only if t is less than or equal to the number of carries in the addition $(n - i) + i$ in base p.*

For a positive integer m and a prime p, we write $p^{t_m} \| m$ if p^{t_m} divides m and $p^{t_m + 1}$ does not; that is, t_m is the greatest integer such that p^{t_m} divides m. Theorem 3.4 is based on the following two lemmas.

Lemma 3.5. *Let n be a positive integer, and let p be a prime. If $p^t \| n!$, then*

$$t = \left\lfloor \frac{n}{p} \right\rfloor + \left\lfloor \frac{n}{p^2} \right\rfloor + \left\lfloor \frac{n}{p^3} \right\rfloor + \cdots .$$

Proof: First we note that this sum is a finite sum, because for large m, $n < p^m$ and $\left\lfloor \frac{n}{p^m} \right\rfloor = 0$. Let m be the smallest integer such that $n < p^m$. It suffices to show that

$$t = \left\lfloor \frac{n}{p} \right\rfloor + \left\lfloor \frac{n}{p^2} \right\rfloor + \cdots + \left\lfloor \frac{n}{p^m} \right\rfloor .$$

For each positive integer i, define t_i such that $p^{t_i} \| i$. Because p is prime, we have $p^{t_1 + t_2 + \cdots + t_n} \| n!$, or $t = t_{n!} = t_1 + t_2 + \cdots + t_n$. On the other hand, $\left\lfloor \frac{n}{p^k} \right\rfloor$ counts all multiples of p^k that are less than or equal to n exactly once. Thus the number $i = p^{t_i} \cdot a$ (with a and p relatively prime) is counted t_i times in the sum

$$\left\lfloor \frac{n}{p} \right\rfloor + \left\lfloor \frac{n}{p^2} \right\rfloor + \cdots + \left\lfloor \frac{n}{p^m} \right\rfloor,$$

namely, in the terms $\left\lfloor \frac{n}{p} \right\rfloor$, $\left\lfloor \frac{n}{p^2} \right\rfloor$, $\ldots$, $\left\lfloor \frac{n}{p^i} \right\rfloor$. Therefore, for each $1 \leq i \leq n$, the number i contributes t_i in both

$$\left\lfloor \frac{n}{p} \right\rfloor + \left\lfloor \frac{n}{p^2} \right\rfloor + \cdots + \left\lfloor \frac{n}{p^m} \right\rfloor$$

and

$$t_1 + t_2 + \cdots + t_n.$$

Hence

$$t = t_1 + t_2 + \cdots + t_n = \left\lfloor \frac{n}{p} \right\rfloor + \left\lfloor \frac{n}{p^2} \right\rfloor + \cdots + \left\lfloor \frac{n}{p^m} \right\rfloor.$$

If you are still not convinced, consider the matrix $\mathbf{M} = (x_{i,j})$ with m rows and n columns, where m is the smallest integer such that $p^m > n$. We define

$$x_{i,j} = \begin{cases} 1 & \text{if } p^i \text{ divides } j, \\ 0 & \text{otherwise.} \end{cases}$$

Then the number of 1's in the jth column of the matrix $\mathbf{M}$ is t_j, implying that the column sums of $\mathbf{M}$ are $t_1, t_2, \ldots, t_n$. Hence the sum of all of the entries in $\mathbf{M}$ is t. On the other hand, the 1's in the ith row denote the numbers that are multiples of p^i. Consequently, the sum of the entries in the ith row is $\left\lfloor \frac{n}{p^i} \right\rfloor$. Thus the sum of the entries in $\mathbf{M}$ is also $\sum_{i=1}^{m} \left\lfloor \frac{n}{p^i} \right\rfloor$. It follows that

$$t = \left\lfloor \frac{n}{p} \right\rfloor + \left\lfloor \frac{n}{p^2} \right\rfloor + \cdots + \left\lfloor \frac{n}{p^m} \right\rfloor,$$

as desired.   ∎

Lemma 3.6. *Let p be a prime, and let n be a positive integer with $n = (\overline{n_m n_{m-1} \ldots n_0})_p$. If $p^t \| n!$, then*

$$t = \frac{n - (n_m + n_{m-1} + \cdots + n_0)}{p - 1}.$$

Proof: By Lemma 3.5, it suffices to show that

$$t = \left\lfloor \frac{n}{p} \right\rfloor + \left\lfloor \frac{n}{p^2} \right\rfloor + \cdots + \left\lfloor \frac{n}{p^m} \right\rfloor = \frac{n - (n_m + n_{m-1} + \cdots + n_0)}{p - 1}.$$

Because

$$n = (\overline{n_m n_{m-1} \ldots n_0})_p = n_0 + n_1 p + \cdots + n_m p^m,$$

we have

$$
\begin{aligned}
\left\lfloor \frac{n}{p} \right\rfloor &= && n_1 + && n_2 p + && n_3 p^2 + && \cdots + && n_m p^{m-1}, \\
\left\lfloor \frac{n}{p^2} \right\rfloor &= && && n_2 + && n_3 p + && \cdots + && n_m p^{m-2}, \\
\cdots && && && \cdots && \cdots && \cdots && \cdots \\
\left\lfloor \frac{n}{p^m} \right\rfloor &= && && && && && && n_m.
\end{aligned}
$$

Adding the left-hand sides of the above equations gives t. Adding the right-hand sides of the above equations by columns yields

$$n_1 + n_2(1 + p) + \cdots + n_m[1 + p + p^2 + \cdots + p^{m-1}]$$
$$= n_1 \cdot \frac{p-1}{p-1} + n_2 \cdot \frac{p^2 - 1}{p-1} + \cdots + n_m \cdot \frac{p^m - 1}{p-1}$$
$$= \frac{1}{p-1} \left[n_1(p-1) + n_2(p^2 - 1) + \cdots + n_m(p^m - 1) \right]$$
$$= \frac{1}{p-1} \left[n_0 + n_1 p + n_2 p^2 + \cdots + n_m p^m - (n_0 + n_1 + \cdots + n_m) \right]$$
$$= \frac{n - (n_0 + n_1 + \cdots + n_m)}{p-1},$$

as desired.   ∎

The proofs for both Lemmas 3.5 and 3.6 are typical examples of Calculating in Two Ways, a technique that we will discuss extensively in Chapter 7. Now we prove Theorem 3.4. We actually prove that the largest nonnegative integer t such that p^t divides $\binom{n}{i}$ is exactly the number of carries in the addition $(n - i) + i$ in base p.

Proof: Let $n! = (\overline{a_m a_{m-1} \ldots a_0})_p$, $i! = (\overline{b_k b_{k-1} \ldots b_0})_p$, $(n-i)! = (\overline{c_\ell c_{\ell-1} \ldots c_0})_p$. Because $1 \le i \le n$, it follows that $k, \ell \le m$. Without loss of generality, we assume that $k \le \ell$. Let a, b, c, and t' be integers such that $p^a \| n!$, $p^b \| i!$, $p^c \| (n-i)!$, and $p^{t'} \| \binom{n}{i}$. Then $t' = a - b - c$.

By Lemma 3.6, we have

$$a = \frac{n - (a_m + a_{m-1} + \cdots + a_0)}{p - 1},$$

$$b = \frac{i - (b_k + b_{k-1} + \cdots + b_0)}{p - 1},$$

$$c = \frac{(n - i) - (c_\ell + c_{\ell-1} + \cdots + c_0)}{p - 1}.$$

Thus,

$$t' = \frac{-(a_m + \cdots + a_0) + (b_k + \cdots + b_0) + (c_\ell + \cdots + c_0)}{p - 1}. \quad (\dagger)$$

On the other hand, if we add $n - i$ and i in base p, we have

$$
\begin{array}{cccccccccc}
 & & & b_k & b_{k-1} & \cdots & & b_1 & b_0 \\
 & & c_\ell & c_{\ell-1} & \cdots & c_k & c_{k-1} & \cdots & c_1 & c_0 \\
\hline
a_m & a_{m-1} & \cdots a_\ell & a_{\ell-1} & \cdots & a_k & a_{k-1} & \cdots & a_1 & a_0.
\end{array}
$$

Then we have either $b_0 + c_0 = a_0$ (with no carry) or $b_0 + c_0 = a_0 + p$ (with a carry of 1). More generally, we have

$$b_0 + c_0 = a_0 + \alpha_1 p,$$

$$b_1 + c_1 + \alpha_1 = a_1 + \alpha_2 p,$$

$$b_2 + c_2 + \alpha_2 = a_2 + \alpha_3 p,$$

$$\cdots\cdots\cdots$$

$$b_m + c_m + \alpha_m = a_m,$$

where α_i denotes the carry at the $(i-1)^{\text{th}}$ digit from the right. (Note also that $b_j = 0$ for $j > k$ and that $c_j = 0$ for $j > \ell$.) Adding the above equations together yields

$$(b_0 + \cdots + b_k) + (c_0 + \cdots + c_\ell) = (a_0 + \cdots + a_m) + (p - 1)(\alpha_1 + \cdots + \alpha_m).$$

Thus, equation ($\dagger$) becomes

$$t' = \alpha_1 + \cdots + \alpha_m,$$

as desired. ∎

The next two examples are applications of Theorems 3.3 and 3.4.

Example 3.13. Let p be a prime. Let n be a positive integer, and let $n = (\overline{n_m n_{m-1} \ldots n_0})_p$ be the base p representation of n. Then:

(a) there are exactly

$$(n_m + 1)(n_{m-1} + 1) \cdots (n_0 + 1)$$

numbers among $\binom{n}{0}, \binom{n}{1}, \ldots, \binom{n}{n}$ that are not divisible by p;

(b) p divides each of $\binom{n}{1}, \binom{n}{2}, \ldots, \binom{n}{n-1}$ if and only if $n = p^k$ for some positive integer k; and

(c) p does not divide any of $\binom{n}{0}, \binom{n}{1}, \ldots, \binom{n}{n}$ if and only if $n = s \cdot p^k - 1$ for some positive integer k and some integer s with $1 \leq s \leq p-1$.

Solution: These results follow directly from Theorem 3.3.

(a) By the congruence relation (**), $\binom{n}{i} \equiv 0 \pmod{p}$ if and only if $i_j > n_j$ for some $0 \leq j \leq m$. Hence $\binom{n}{i}$ is not divisible by p for exactly those $i = i_0 + i_1 p + \cdots + i_m p^m$ with $0 \leq i_j \leq n_j$ for $0 \leq j \leq m$.

(b) If $n = p^k$ for some positive integer k, then $n = (\overline{10 \ldots 0})_p$ (with k 0's). For $1 \leq i \leq n - 1$, write $i = i_0 + i_1 p + \cdots + i_m p^m$, where $i_m = 0$ and $0 \leq i_0, i_1, \ldots, i_{m-1} \leq p - 1$. Then i_j is positive for some $0 \leq j \leq m-1$, so $\binom{n_j}{i_j} = \binom{0}{i_j} = 0$. By Theorem 3.3, $\binom{n}{i} \equiv 0 \pmod{p}$.

   If $n \neq p^k$ for positive integers k and $n_m > 1$, then letting $i = (\overline{1 n_{m-1} \ldots n_0})_p$, we have $1 \leq i \leq n - 1$ and $\binom{n}{i} \equiv \binom{n_m}{1} \not\equiv 0 \pmod{p}$.

   If $n \neq p^k$ for positive integers k and $n_m = 1$, then $n_j > 0$ for some $0 \leq j \leq m - 1$. Letting $i = (\overline{n_j \ldots n_0})_p$ gives $\binom{n}{i} \not\equiv 0 \pmod{p}$.

(c) If $n = s \cdot p^k - 1$ for some positive integer k, then $n = (\overline{(s-1)(p-1) \ldots (p-1)})_p$ (with k $(p-1)$'s). For $0 \leq i \leq n$, write $i = i_0 + i_1 p + \cdots + i_m p^m$, where $0 \leq i_0, i_1, \ldots, i_{m-1} \leq p - 1$ and $0 \leq i_m \leq s - 1$. Because p is prime, p does not divide the numerator of either $\frac{(p-1)!}{i_j!(p-1-i_j)!}$ or $\frac{(s-1)!}{i_m!(s-1-i_m)!}$; that is, p does not divide either $\binom{p-1}{i_j}$ or $\binom{s-1}{i_m}$. By Theorem 3.3, p does not divide $\binom{n}{i}$ for $1 \leq i \leq n$.

   If n cannot be written in the form $s \cdot p^k - 1$, then $n_j < p - 1$ for some $0 \leq j \leq m - 1$. Letting $i = (\overline{(p-1)0 \ldots 0})_p$ ($j - 1$ 0's), we have $\binom{n}{i} \equiv 0 \pmod{p}$ by Theorem 3.3. ∎

Setting $p = 2$ in parts (b) and (c) of Example 3.13, we conclude that all of the entries in the $(2^k - 1)^{\text{th}}$ row of Pascal's triangle are odd and that all of the entries (except for the two ends) in the $2^{k\text{th}}$ row are even.

Example 3.14. Determine all positive integers k satisfying the following property: There exists a positive integer $n > 1$ (which depends on k) such that the binomial coefficient $\binom{n}{i}$ is divisible by k for any $1 \le i \le n - 1$.

There is also a much easier version of this problem:

> *Prove or disprove the following claim: For any positive integer k, there exists a positive integer $n > 1$ such that the binomial coefficient $\binom{n}{i}$ is divisible by k for any $1 \le i \le n-1$.*

This problem appeared in the Hungary-Israel Bi-national Mathematical Competition in 1999. The solution to Example 3.14 is due to Reid Barton, the only four-time IMO gold medalist in the world as of 2003.

Solution: The claim is obviously true for $k = 1$; we prove that the set of positive integers $k > 1$ for which the claim holds is exactly the set of primes. By Example 3.13 (b), the claim holds for a prime k exactly when $n = k^m$.

Now suppose the claim holds for some $k > 1$ with the number n. If some prime p divides k, the claim must also hold for p with the number n. Thus n must equal a prime power p^m where $m \ge 1$. Then $k = p^r$ for some $r \ge 1$ as well, because if two primes p and q divided k, then n would equal a perfect power of both p and q, which is impossible.

Choose $i = p^{m-1}$. Kummer's Theorem states that $p^t \mid \binom{n}{i}$ if and only if t is less than or equal to the number of carries in the addition $(n - i) + i$ in base p. There is only one such carry, which occurs between the p^{m-1} and p^m places:

$$
\begin{array}{r}
1 \quad 0 \quad 0 \ \ldots \ 0 \\
+ \quad p-1 \quad 0 \quad 0 \ \ldots \ 0 \\
\hline
1 \quad\quad 0 \quad 0 \quad 0 \ \ldots \ 0
\end{array}
$$

Thus we must have $r \le 1$, and k must be prime, as claimed. ∎

(Alternatively, for $n = p^m$ and $i = p^{m-1}$ we have

$$
\binom{n}{i} = \prod_{j=0}^{p^{m-1}-1} \frac{p^m - j}{p^{m-1} - j}.
$$

When $j = 0$, then $\frac{p^m-j}{p^{m-1}-j} = p$. Otherwise, $0 < j < p^{m-1}$, so that if $p^t < p^{m-1}$ is the highest power of p dividing j, then it is also the highest power of p dividing both $p^m - j$ and $p^{m-1} - j$. Therefore, $\frac{p^m-j}{p^{m-1}-j}$ contributes one factor of p to $\binom{n}{i}$ when $j = 0$ and zero factors of p when $j > 0$. Thus p^2 does not divide $\binom{n}{i}$, and again $r \leq 1$.)

We close this section with the following generalization of Theorem 3.1.

Theorem 3.7.   *Let m and n be positive integers. Then*

$$(x_1 + x_2 + \cdots + x_m)^n$$

$$= \sum_{\substack{n_1,n_2,\ldots,n_m \geq 0 \\ n_1+n_2+\cdots+n_m=n}} \binom{n}{n_1, n_2, \ldots, n_m} x_1^{n_1} x_2^{n_2} \cdots x_m^{n_m}.$$

The proof of the theorem is similar to that of Theorem 3.1. We leave it as an exercise.

Exercises 3

3.1. Let n, m, k be nonnegative integers such that $m \leq n$. Show that

$$\binom{n}{k}\binom{k}{m} = \binom{n}{m}\binom{n-m}{k-m}.$$

3.2. Prove Theorem 3.7.

3.3. What is the value of the constant term in the expansion of $\left(x^2 + \frac{1}{x^2} - 2\right)^{10}$?

3.4. Let n be a positive integer. Prove that

$$\sum_{k=0}^{n} k^2 \binom{n}{k} = n(n+1)2^{n-2} \quad \text{and} \quad \sum_{k=0}^{n} \frac{(-1)^k}{k+1}\binom{n}{k} = \frac{1}{n+1}.$$

3.5. Let n be a nonnegative integer. Prove that

$$\sum_{i=0}^{k}(-1)^i \binom{n}{i} = (-1)^k \binom{n-1}{k}.$$

3.6. Let n be an odd positive integer. Prove that the array

$$\binom{n}{0}, \binom{n}{1}, \ldots, \binom{n}{\frac{n-1}{2}}$$

contains an even number of odd numbers.

3.7. Let n be a positive integer. Determine the number of odd numbers in $\binom{n}{0}, \binom{n}{1}, \ldots, \binom{n}{n}$.

3.8. Prove that the binomial coefficients $\binom{2^n}{k}$, $k = 1, 2, \ldots, 2^{n-1} - 1, 2^{n-1} + 1, \ldots, 2^n - 1$, are all divisible by 4.

3.9. Let p be any prime number. Prove that

$$\binom{2p}{p} \equiv 2 \pmod{p^2}.$$

3.10. Let n be a nonnegative integer. Show that

$$\sum_{k=0}^{n} k \binom{n}{k}^2 = n \binom{2n-1}{n-1}.$$

3.11. [MOSP 2001, Cecil Rousseau] Let a_n denote the number of nonempty sets S such that

(i) $S \subseteq \{1, 2, \ldots, n\}$;

(ii) all elements of S have the same parity;

(iii) each element $k \in S$ satisfies $k \geq 2|S|$, where $|S|$ denotes the number of elements in S.

Prove that

$$a_{2m-1} = 2(F_{m+1} - 1) \quad \text{and} \quad a_{2m} = F_{m+3} - 2$$

for all $m \geq 1$, where F_n is the nth Fibonacci number.

4
Bijections

In this chapter, we introduce a fundamental technique in solving combinatorial problems.

In order to count the elements of a certain set, we replace them with those of another set that has the same number of elements and whose elements are more easily counted.

The following theorem is rather intuitive.

Theorem 4.1. *Let A and B be finite sets, and let f be an injective function from A to B. Then there are at least as many elements in B as in A. Furthermore, if f is bijective, then A and B have the same number of elements.*

Proof: Let $A = \{a_1, a_2, \ldots, a_n\}$ for some positive integer n. Because f is bijective, $f(a_1), f(a_2), \ldots, f(a_n)$ are all distinct. Hence B has at least n elements. If f is also surjective, then each element in B is the image of some element in A. Hence $B = \{f(a_1), f(a_2), \ldots, f(a_n)\}$, implying that A and B have the same number of elements. ∎

Example 4.1. Each of the vertices of a regular nonagon has been colored either red or blue. Prove that there exist two congruent monochromatic triangles; that is, triangles whose vertices are all the same color.

Solution: We call a monochromatic triangle red (blue) if all of its vertices are red (blue). Because there are nine vertices colored in two colors, at least five must be of the same color. Without loss of generality, we say that this color is red. Hence there are at least $\binom{5}{3} = 10$ red triangles. We now prove that there are two congruent red triangles.

Let $A_1, A_2, \cdots, A_9$ denote the vertices of the nonagon (Figure 4.1), and let ω be the its circumcircle. The vertices of the nonagon cut ω into nine equal arcs. We call each of these nine arcs a *piece*. Let $A_i A_j A_k$ be a triangle with $A_i A_j \leq A_j A_k \leq A_k A_i$. Denote by $a_{i,j}$ the number of pieces in the arc $\widehat{A_i A_j}$, not containing the point A_k, and define $a_{j,k}$ and $a_{k,i}$ analogously. Then define a map that maps the triangle $A_i A_j A_k$ to the triple $(a_{i,j}, a_{j,k}, a_{k,i})$. It is clear that $1 \leq a_{i,j} \leq a_{j,k} \leq a_{k,i} \leq 7$ and $a_{i,j} + a_{j,k} + a_{k,i} = 9$. For example, the triangle with vertices A_2, A_4, A_8 is read as triangle $A_4 A_2 A_8$ and mapped to the triple $(2, 3, 4)$.

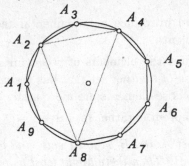

Figure 4.1.

Congruent triangles map to the same triple, while incongruent triangles map to distinct triples. Hence we have built a bijection between the classes of congruent triangles and the set of ordered triples of positive integers (a, b, c) with $a \leq b \leq c$ and $a + b + c = 9$. It is not difficult to list all such triples: $(1, 1, 7)$, $(1, 2, 6)$, $(1, 3, 5)$, $(1, 4, 4)$, $(2, 2, 5)$, $(2, 3, 4)$, $(3, 3, 3)$. Hence there are seven classes of congruent triangles. Since there are at least 10 red triangles, some class must contain at least two red triangles and hence there are at least two congruent red triangles.   ∎

Example 4.2. Let n be a positive integer. In how many ways can one write a sum of (at least two) positive integers that add up to n? Consider the same set of integers written in a different order as being different. (For example, there are 3 ways to express 3 as $3 = 1 + 1 + 1 = 2 + 1 = 1 + 2$.)

Solution: Consider the $n-1$ spaces between the n 1's in the following arrangement:

$$\underbrace{(1\_\_1\_\_\cdots\_\_1)}_{n\ 1's}.$$

Define two possible states, 0 and 1, for each space. If the space is in state 0, we put a " $+$ " there. If the space is in state 1, we put a ") $+$ (" there. There are 2^{n-1} such $(n-1)$-digit binary sequences, and each of these sequences uniquely corresponds to a writing of n as a sum of ordered positive integers and vice versa. This is because by the order of operations, we need to add the numbers (here the 1's) between each pair of parentheses first. For example, for $n=8$, the sequence 0101100 corresponds to

$$(1+1)+(1+1)+(1)+(1+1+1)=2+2+1+3=8.$$

Among these binary sequences, only one does not correspond to a legal sum: the sequence $00\ldots0$, because it leads to $(1+1\cdots+1)=n$. Hence there are $2^{n-1}-1$ ways to express n as a sum of ordered positive integers. ∎

A much more difficult and interesting question is to find the number of ways in which one can write a sum of (at least two) integers that add up to n, if different orders of the summands are not distinguished. We will discuss this problem in Chapter 8.

Example 4.3.   [AHSME 1992] Ten points are selected on the positive x-axis, $\mathbf{X}^+$, and five points are selected on the positive y-axis, $\mathbf{Y}^+$. The fifty segments connecting the ten points on $\mathbf{X}^+$ to the five points on $\mathbf{Y}^+$ are drawn. What is the maximum possible number of points of intersection of these fifty segments in the interior of the first quadrant? (See Figure 4.2.)

Solution: A point of intersection in the first quadrant is obtained whenever two of the segments cross to form an ×. An × is uniquely determined by selecting two of the points on $\mathbf{X}^+$ and two of the points on $\mathbf{Y}^+$.

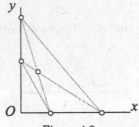

Figure 4.2.

The maximum number of these intersections is obtained by selecting the points of $\mathbf{X}^+$ and $\mathbf{Y}^+$ so that no three of the fifty segments intersect at the same point. Therefore, the maximum number of intersections is $\binom{10}{2}\binom{5}{2} = 45 \cdot 10 = 450$.   ∎

Example 4.4. [China 1991, by Weichao Wu] Let n be an integer with $n \geq 2$, and define the sequence $S = (1, 2, \ldots, n)$. A subsequence of S is called arithmetic if it has at least two terms and it is an arithmetic progression. An arithmetic subsequence is called maximal if this progression cannot be lengthened by the inclusion of another element of S. Determine the number of maximal arithmetic subsequences.

Solution: First we consider the case when n is even. Assume that $n = 2m$ for some positive integer m. Let $a_1 < a_2 < \cdots < a_k$ be a maximal arithmetic subsequence for some integer $k \geq 2$. Then $a_1 \leq m$. Otherwise, $a_1 \geq m + 1$ and $a_2 - a_1 \leq 2m - (m+1) = m - 1$. We can then add $2a_1 - a_2$ to lengthen the subsequence, because $2a_1 - a_2 = a_1 - (a_2 - a_1) \geq m + 1 - (m - 1) = 2$, contradicting the maximality of the subsequence. Likewise, we can show that $a_2 \geq m + 1$. Hence a maximal arithmetic subsequence must have two consecutive terms a_i and a_{i+1} with $1 \leq a_i \leq m < m + 1 \leq a_{i+1} \leq n$. We map the subsequence to the pair of consecutive terms (a_i, a_{i+1}). Such a pair of consecutive terms clearly determines the common difference of the subsequence. The maximality of the subsequence means that all other terms are uniquely determined (just use the common difference to extend the sequence to the left and to the right until it cannot be extended anymore within the range between 1 and n). Hence the whole subsequence is determined by the pair of consecutive terms, and our map is a one-to-one map; that is, a bijection. It is clear that there are m possible values for a_i, and m

possible values for a_{i+1}, implying that there are m^2 pairs of numbers (a_i, a_{i+1}), and m^2 maximal arithmetic subsequences in this case.

In a similar way, we can show that there are $m(m + 1)$ maximal arithmetic subsequences if $n = 2m + 1$ for some positive integer m.

Putting the above together, we conclude that there are $\left\lfloor \frac{n^2}{4} \right\rfloor$ maximal arithmetic subsequences for any given n. ∎

Bijections can sometimes be used between sets that are not necessarily finite. Here is an example.

Example 4.5. [PEA Math Materials] Suppose that two PEA administrators are the only persons who have dial-in access to the Academy's Internet fileserver, which can handle only one call at a time. Each has a 15-minute project to do and hopes to use the fileserver between 4 PM and 6 PM today. Neither will call after 5:45 PM, and neither will call more than once. At least one of them will succeed. What is the probability that they both complete their tasks?

Solution: Let Adrian and Zachary be the two administrators. Each can dial–in in a 105-minute window. If Adrian calls x minutes after 4 PM and Zachary calls y minutes after 4 PM, we can map this pair of calling times to the point (x, y) in the coordinate plane (Figure 4.3). This is certainly a bijection.

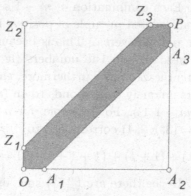

Figure 4.3.

The shaded region is the set of points (x, y) with $|x - y| \le 15$. One of them cannot complete his task if and only if he calls while the other is still on the line, namely when $|x - y| \le 15$. Hence, the probability that both of them complete their tasks successfully is the ratio of the area of the unshaded region to the area of the whole square. Thus the answer to the problem is $\left(\frac{90}{105}\right)^2 = \frac{36}{49}$. ■

The following theorem is very useful in solving combinatorial problems.

Theorem 4.2. *Let m and n be positive integers.*

(a) There are $\binom{n-1}{m-1}$ ordered m-tuples $(x_1, x_2, \ldots, x_m)$ of positive integers satisfying the equation $x_1 + x_2 + \cdots + x_m = n$.

(b) There are $\binom{n+m-1}{m-1}$ ordered m-tuples $(x_1, x_2, \ldots, x_m)$ of nonnegative integers satisfying the equation $x_1 + x_2 + \cdots + x_m = n$.

Proof: There are many proofs. We choose one that is similar to the solution to Example 4.2.

(a) As in the solution of Example 4.2, we consider the $n-1$ spaces between the n 1's in the following arrangement:

$$\underbrace{(1\underline{\quad}1\underline{\quad}\cdots\underline{\quad}1)}_{n\ 1\text{'s}}.$$

Define two possible states, 0 and 1, for each space. If the space is in state 0, we put a "$+$" there. If the space is in state 1, we put a *separator*, ")$+$(" there. We choose $m-1$ of these spaces and put a separator in each. In other words, we consider the $(n-1)$-digit binary sequences containing exactly $m-1$ 1's. There are $\binom{n-1}{m-1}$ such sequences. Each combination of $m-1$ separators uniquely corresponds to an ordered m-tuple of positive integers with their sum equal to n and vice versa. This is because by the order of operations, we need to add the numbers (here the 1's) within each pair of parentheses first. Furthermore, each combination of $m-1$ separators uniquely corresponds to an $(n-1)$-digit binary sequence with $m-1$ 1's. For example, for $n = 10$ and $m = 4$, $(x_1, x_2, x_3, x_4) = (3, 1, 2, 4)$ corresponds to

$$(1+1+1) + (1) + (1+1) + (1+1+1+1) = 3+1+2+4 = 10$$

and 001101000. Hence there are $\binom{n-1}{m-1}$ such ordered m-tuples.

(b) We need to modify our method, because x_i can be 0, which means that separators are allowed to be next to each other. We consider $n + m - 1$ circles in a row. We pick any $m - 1$ of these and make them separators, then label the chosen circles $s_1, s_2, \ldots, s_{m-1}$ from left to right. For $1 \leq i \leq m - 2$, let x_{i+1}

be the number of circles that are not chosen between s_i and s_{i+1}. Let x_1 be the number of circles to the left of s_1, and let x_m be the number of circles to the right of s_{m-1}. It is not difficult to see that this is a bijection. For example, for $m = 6$ and $n = 4$, $(x_1, x_2, x_3, x_4, x_5, x_6) = (1, 0, 0, 1, 0, 2)$ corresponds to the diagram of Figure 4.4.

Figure 4.4.

Hence, there are $\binom{n+m-1}{m-1}$ such ordered m-tuples satisfying the conditions of the problem.   ∎

Indeed, there is also a bijection between the m-tuples in Theorem 4.2 (a) and (b). An m-tuple $(x_1, x_2, \ldots, x_m)$ of positive integers with $x_1 + x_2 + \cdots + x_m = n$ can be uniquely mapped to an m-tuple $(y_1, y_2, \ldots, y_m)$ of nonnegative integers with $y_1 + y_2 + \cdots + y_m = n - m$ under the bijective mapping $(y_1, y_2, \ldots, y_m) = (x_1 - 1, x_2 - 1, \ldots, x_m - 1)$.

Let us see a few examples that put Theorem 4.2 into action.

Example 4.6.   Five regular dice are rolled. What is the probability that the sum of the five numbers shown is equal to 14?

Solution: Let $d_1, d_2, \ldots, d_5$ denote the dice, and let x_i be the number shown on d_i. Each x_i can take six possible values. Hence there are 6^5 total possible outcomes. Let A be the set of all outcomes with sums equal to 14. We need to calculate $\frac{|A|}{6^5}$. Hence, we need to count the number of 5-tuples $(x_1, x_2, \ldots, x_5)$ in integers such that $1 \le x_i \le 6$ and $x_1 + x_2 + \cdots + x_5 = 14$. Without the restriction $x_i \le 6$, by Theorem 4.2.(a) there are $\binom{14-1}{5-1} = \binom{13}{4} = 715$ such 5-tuples.

A 5-tuple is bad if $x_i > 6$ for some $1 \le i \le 5$. For $1 \le i \le 5$, let B_i be the set of bad 5-tuples with $x_i > 6$. It is clear that B_i and B_j are exclusive for $i \ne j$ (otherwise $x_1 + \cdots + x_5 > 6 + 6 + 1 + 1 + 1 = 15$). By symmetry, we also have $|B_i| = |B_j|$. Hence, there are $5|B_1|$ bad 5-tuples. Map $(x_1, x_2, \ldots, x_5) \in B_1$ to $(y_1, y_2, \ldots, y_5)$ with $y_1 = x_1 - 6$ and $y_i = x_i$ for $2 \le i \le 5$. It is clear that this map is a bijection between B_1 and the set of 5-tuples $(y_1, y_2, \ldots, y_5)$ in positive integers with $y_1 + y_2 + \cdots + y_5 = 8$. Hence, $|B_1| = \binom{8-1}{5-1} = \binom{7}{4} = 35$ and $5|B_1| = 175$.

Therefore, $|A| = 715 - 175 = 540$, and the answer to the problem is $\frac{540}{6^5} = \frac{5}{72}$.   ∎

The number 14 was chosen carefully in the last problem to make the problem more tractable. If we replace 14 by, say, 17, then we would have had many more cases to consider, because there could have been two numbers greater than 6. A more powerful way of dealing with this kind of problem will be discussed in Chapter 8.

Example 4.7. [AIME 2000] Given eight distinguishable rings, find the number of possible five-ring arrangements on the four fingers (not the thumb) of one hand. (The order of the rings on each finger is significant, but it is not required that each finger have a ring.)

Solution: There are $\binom{8}{5}$ ways to select the rings to be worn. There are $\binom{8}{5}$ ways to assign 5 indistinguishable rings to 4 fingers, which can be seen by inserting 3 separators into a row of five rings to designate the spaces between the four fingers. In other words, if a, b, c, d are the numbers of rings on the fingers, we need to find the number of ordered quadruples (a, b, c, d) of nonnegative integers such that $a + b + c + d = 5$. By Theorem 4.2 (b), there are $\binom{8}{5}$ such quadruples. For each assignment of 5 indistinguishable rings, there are 5! assignments of distinguishable rings. Thus the desired answer is

$$\binom{8}{5}\binom{8}{5}5! = 376320.$$

∎

Example 4.8. The Security Bank of the Republic of Fatland has 15 senior executive officers. Each officer has an access card to the bank's vault. There are m distinct codes stored in the magnetic strip of each access card. To open the vault, each officer who is present puts his access card in the vault's electronic lock. The computer system then collects all of the distinct codes on the cards, and the vault is unlocked if and only if the set of the codes matches the set of n (distinct) preassigned codes. For security reasons, the bank's vault can be opened if and only if at least six of the senior officers are present. Find the values of n and m such that n is minimal and the vault's security policy can be achieved. (The elements in a set have no order.)

Solution: For every five officers, there is at least one code missing from their cards. Let A be the set of groups of five officers, and let B be the set of n codes needed to open the vault. Map every group in A to one of the codes missing from the cards in the group. It is possible that there is more than one such code. We will take care of such situations when we focus on the minimality of n. For now, we just defined a map, denoted by f. We claim that f is injective. We prove our claim indirectly. Assume not; then for some distinct elements a_1 and a_2 in A, $f(a_1) = f(a_2) = c$. This means that the two 5-officer groups a_1 and a_2 are missing the same code c from their cards. Because a_1 and a_2 are distinct, there are at least six officers in the union of a_1 and a_2. But when these six officers are present, they are still missing the code c, so they cannot open the vault. Thus our assumption is wrong, and f is injective. Hence, by Theorem 4.1, $n = |B| \geq |A| = \binom{15}{5} = 3003$.

Now we will show that $n = 3003$ is enough. Assign a different code $c(a)$ for every 5-officer group a. This is possible since there are $\binom{2002}{5} = 3003$ different 5-officer groups. Write the code $c(a)$ on the card of every officer that is not in the group a. Each officer will have $\binom{14}{5} = 2002$ codes on his or her access card. Any group s of 6 officers will have all codes on their cards, since for each code $f(g)$ defined by a 5-officer group g there will be at least one officer in s that is not in g and such an officer will have the code $f(g)$ on his or her access card.

Thus $n = 3003$ is the minimal possible number of codes, and the corresponding value for m is 2002,   ∎

One can also obtain the value of m in the following way. Each of the $\binom{15}{5}$ codes in B will be written 10 times. Hence there are a total of $10\binom{15}{5}$ codes written on all of the cards, with each card having m codes written on it, so $m = \frac{10}{15}\binom{15}{5}$. Putting the two results together, we have $\binom{14}{5} = \frac{10}{15}\binom{15}{5}$, which is a special case of Theorem 3.2 (d). Thus we used a complicated combinatorial model to prove a simple combinatorial identity. What's the big deal? Well, the idea of calculating a certain quantity in two ways is fundamental in problem solving. We all know how important it is to know how to set up and solve equations in algebra.

Example 4.9. [AIME 2001, by Richard Parris] The numbers 1, 2, 3, 4, 5, 6, 7, and 8 are randomly written on the faces of a regular octahedron so that each face contains a different number. Compute the probability that no two consecutive numbers, where 8 and 1 are considered to be consecutive, are written on faces that share an edge.

Solution: It is helpful to consider the cube $ABCDEFGH$ shown in Figure 4.5. The cube is called the *dual* of the octahedron. The vertices of the cube represent the faces of the dotted octahedron, and the edges of the cube represent adjacent octahedral faces. Each assignment of the numbers 1, 2, 3, 4, 5, 6, 7, and 8 to the faces of the octahedron corresponds to a permutation of $ABCDEFGH$, and thus to an octagonal circuit of these vertices. We look for the number of circuits such that no adjacent vertices in the circuits (with the first and last vertices in the circuits being considered adjacent) are the end points of an edge of the cube; that is, we look for the number of circuits around the vertices of the cube that travel only through the (face or internal) diagonals of the cube.

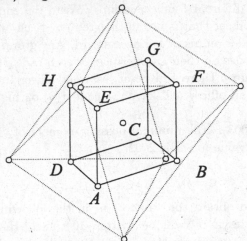

Figure 4.5.

We split the vertices into the two groups, $G_1 = \{A, C, F, H\}$ and $G_2 = \{G, E, D, B\}$. In this way, for any vertex in any of the groups, the vertices that are not allowed to be adjacent to it are in the other group.

Let $a_1 a_2 \ldots a_8$ be an acceptable circuit. Its leading vertex a_1 can be chosen either from G_1 or G_2. Without loss of generality, we assume that a_1 is from G_1. Let a_i, a_j, a_k, $2 \leq i < j < k \leq 8$, be the other three vertices in G_1. Without loss of generality, we assume that $(a_1, a_i, a_j, a_k) = (A, C, F, H)$, since, by symmetry, any of the other 23 permutations of the elements of G_1 yields the same number of possibilities. It suffices to count the number of ways of inserting the elements of G_2 into the spaces in the sequence

$$A\_\_C\_\_F\_\_H\_\_ .$$

(More than one element is allowed to be inserted into one space.) Note that the only vertex that is allowed to be inserted after A is G. Similarly, only E can be inserted after C, only D after F, and only B after H. Hence, the required sequences can be categorized into four types:

- *Type I:* The circuit starts with AG. Then ECF must be part of the circuit. Note also that no vertices can be inserted after H (because A is not allowed to be adjacent to neither B nor D). There are three possible circuits of this type:

  $$AGBDECFH, \quad AGDBECFH, \quad \text{and} \quad AGECFDBH.$$

- *Type II:* The circuit starts with ACE. Then DFH must be part of the circuit. There are three possible circuits of this type:

  $$ACEBGDFH, \quad ACEGBDFH, \quad \text{and} \quad ACEDFHBG.$$

- *Type III:* The circuit starts with $ACFD$. Then BH must be part of the circuit. There are two possible circuits of this type:

  $$ACFDEGBH \quad \text{and} \quad ACFDGEBH.$$

- *Type IV:* The circuit starts with $ACFHB$. There are two possible circuits of this type:

  $$ACFHBDEG \quad \text{and} \quad ACFHBEDG.$$

Hence there are 10 circuits of the form $A\_\_C\_\_F\_\_H\_\_$ and there are $2 \times 24 \times 10 = 480$ circuits formed exclusively from cube diagonals. Thus the probability of randomly choosing such a permutation is $\frac{480}{8!} = \frac{1}{84}$. ∎

Rick wrote this problem while teaching his group theory course at PEA. In that course, many symmetry groups are introduced through solving the Rubick's Cube. Readers that are interested in the theoretic background of this problem should read more on theorems of Burnside and Pólya on counting in the presence of symmetries.

Example 4.10. A triangular grid is obtained by tiling an equilateral triangle of side length n by n^2 equilateral triangles of side length 1 (Figure 4.6). Determine the number of parallelograms bounded by line segments of the grid.

Solution: We partition all of the parallelograms into three sets. The sides of the parallelograms must be parallel to exactly two sides of the triangle. Define the set S_{YZ} as the set of those parallelograms whose sides are parallel to XY and ZX, and define the sets S_{ZX} and S_{XY} analogously. By symmetry, we have $|S_{XY}| = |S_{YZ}| = |S_{ZX}|$, and the answer to the problem is $3|S_{YZ}|$.

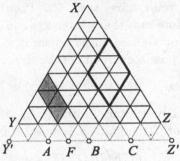

Figure 4.6.

Extend $\overline{XY}$ and $\overline{XZ}$ through Y and Z, respectively, by 1 unit to Y' and Z', respectively. Let p be a parallelogram in S_{YZ}. We extend all four sides of p to intersect the line $Y'Z'$. It is not difficult to see that all four intersections are distinct points on $\overline{Y'Z'}$, and that any four distinct points on $\overline{Y'Z'}$ determine a parallelogram p in S_{YZ}. This is because the vertices of p will be determined by the lines parallel to lines XY and XZ passing through the four points. (For example, the shaded parallelogram corresponds to $\{Y', A, F, B\}$, while the parallelogram with thickened sides corresponds to $\{A, B, C, Z'\}$.) Hence we have a bijection between S_{YZ} and the set of combinations of four distinct points on $\overline{Y'Z'}$. Thus the answer to the problem is $3\binom{n+2}{4}$. ∎

Example 4.11. Bart works at the IOKA movie theater, which has a capacity of 200 seats. On the opening night of Star Wars Episode I: The Phantom Menace, 200 people are standing in line to buy tickets for the movie. The cost of each ticket is \$5. Among the 200 people buying tickets, 100 of them have a single \$5 bill, and 100 of them have a single \$10 bill. Bart, being careless, finds himself with no change at all. The 200 people are in random order in line, and no one is willing to wait for change when they buy their ticket. What is the probability that Bart will be able to sell all of the tickets successfully?

Solution: It might be easier to consider n instead of 100. We consider the people as being indistinguishable. Hence we consider arrangements of n 5's and n 10's. (Otherwise, we just multiply $(n!)^2$ to both the denominator and numerator, which does not affect the value of the desired probability.) There are $\binom{2n}{n}$ ways to arrange the numbers without restriction. Bart is successful if and only if he has received at least as many 5's as 10's at each stage. We arrange the $2n$ numbers in a row and label their places from left to right as $1^{st}, 2^{nd}, \ldots, 2n^{th}$. For $1 \leq i \leq 2n$, let a_i (b_i) be the number of 5's (10's) on or before the ith place. We need to count the number of arrangements such that $a_i \geq b_i$ for $1 \leq i \leq 2n$. Let S denote the set of such arrangements. Let T be the set of all other arrangements. We will find $|T|$ first. Then $|S| = \binom{2n}{n} - |T|$.

For an element $t = (t_1, t_2, \ldots, t_{2n})$ of T, there is an i, $1 \leq i \leq 2n$, such that $a_i < b_i$. Let $f(t)$ denote the first such index. By our definitions, $a_{f(t)} = \frac{f(t)-1}{2}$, $b_{f(t)} = \frac{f(t)+1}{2}$, and $t_{f(t)} = 10$. Thus there is one more 10 than there are 5's up to the $f(t)^{th}$ place. We change all of the 5's to 10's and all of the 10's to 5's after the $f(t)^{th}$ place. Then we obtain a permutation $g(t)$ of $n+1$ 10's and $n-1$ 5's. For example, for $n = 6$ and

$$t = (5, 10, 10, 5, 10, 10, 10, 5, 10, 5, 5, 5),$$

we have

$$(a_1, a_2, \ldots, a_{12}) = (1, 1, 1, 2, 2, 2, 2, 3, 3, 4, 5, 6),$$

$$(b_1, b_2, \ldots, b_{12}) = (0, 1, 2, 2, 3, 4, 5, 5, 6, 6, 6, 6),$$

$$f(t) = 3, \quad a_2 = 1, \quad b_2 = 1, \quad t_3 = 10,$$

$$g(t) = (5, 10, 10, 10, 5, 5, 5, 10, 5, 10, 10, 10).$$

Hence we have defined a map g from T to the set U of permutations of $n+1$ 10's and $n-1$ 5's. We claim that g is bijective.

First we show that g is surjective. Let u be an element in U. We count the appearances of 5's and 10's from left to right. Let i be the smallest integer such that there are more 10's than 5's up to the ith place (from the left). Because there are more 10's than 5's in u, such a place always exists. We change all of the 5's to 10's and all of the 10's to 5's after that place. There are two more 10's than 5's in u. There is one more 10 than there are 5's in the first i places. There is one more 10 than there are 5's in the other $2n - i$ places. After we switch the roles of 5's and 10's, we have exactly the same number of 5's and 10's in the new sequence. It is not difficult to see that the new sequence belongs to T, so g is surjective.

Now we show that g is injective. Let t and t' be distinct elements in T. Assume that the ith, $1 \leq i \leq 2n$, place is the first place in which they are different from each other. Without loss of generality, assume that $t_i = 5$ and $t'_i = 10$. Then $f(t) \neq i$. If $f(t) < i$, then $f(t') = f(t)$, because the first $i - 1$ places of t and t' are the same. Then the ith places of $g(t)$ and $g(t')$ are 10 and 5, respectively, so $g(t) \neq g(t')$. We instead assume that $f(t) > i$. Then the ith place of $g(t)$ is t_i, which is 5. Because the first $i - 1$ places of t and t' are the same, $f(t') \geq i$. Hence the ith place of $g(t')$ is t'_i, which is 10. Hence $g(t) \neq g(t')$. We conclude that $g(t) \neq g(t')$ in all cases; that is, g is injective.

Therefore, $|T| = |U| = \binom{2n}{n-1}$ by Theorem 4.1. Hence,

$$|S| = \binom{2n}{n} - |T| = \binom{2n}{n} - \binom{2n}{n-1} = \frac{1}{n+1}\binom{2n}{n}$$

by Theorem 3.2 (e). Setting $n = 200$ gives $\frac{1}{201}$, the probability that Bart will be able to sell all of the tickets successfully. ∎

This is one of the combinatorial models of the famous **Catalan numbers**, which will be discussed carefully in Chapter 8. In the solution, we took advantage of the first place in which $a_i < b_i$. This application of the well-ordering principle is very useful in combinatorial arguments.

Example 4.12. Let n be a positive integer satisfying the following property: If n dominoes are placed on a 6×6 chessboard with each domino covering exactly two unit squares, then one can always place one more domino on the board without moving any other dominoes. Determine the maximum value of n.

Solution: We claim that the maximum value of n is 11. Figure 4.7 shows an arrangement of 12 dominoes in which one cannot place any more dominoes on the board. Consequently, $n \leq 11$.

Figure 4.7.

It suffices to show that one can always place one more domino if there are 11 dominoes on the board. We approach the problem indirectly by assuming that there is an arrangement of 11 dominoes such that one cannot place another domino on the board. Hence there are $36 - 22 = 14$ unit squares not covered by the dominoes.

Let S_1 denote the upper 5×6 sub-board of the given chessboard. Let A denote the set of unit squares in S_1 that are not covered by the dominoes. Let S_2 denote the bottom row of the chessboard. Because we cannot place one more domino on the board, at least one of any two neighboring unit squares (unit squares sharing an edge) is covered by a domino. Hence there are at least three unit squares in S_2 that are covered by dominoes, so there are at most three uncovered unit squares in S_2. Thus there are at least 11 non-covered unit squares in S_1; that is, $|A| \geq 11$.

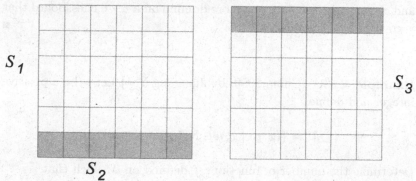

Figure 4.8.

Let S_3 denote the lower 5×6 sub-board of the given chessboard. Let B denote the set of dominoes that lie (completely) in S_3. We will define a map f from A to B. By the definition of S_1, there is always a unit square t directly below a unit square s in S_1. Further, assume that s is in A. Then t must be covered by a domino d in S_3, because otherwise we could place one more domino covering s and t. We define $f(s) = d$. We claim that f is injective. If not, then there are s_1 and s_2 in A with $f(s_1) = f(s_2) = d$. This means that d covers the unit squares directly below s_1 and s_2. Hence s_1 and s_2 must be neighbors. (See Figure 4.9.) But then we can place an extra domino on s_1 and s_2, which is a contradiction. Thus f is indeed injective, and by Theorem 4.1, $|A| \leq |B|$.

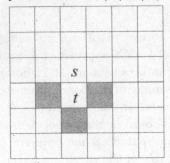 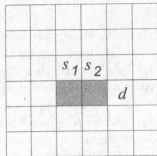

Figure 4.9.

It follows that $|B| \geq 11$. But there are only 11 dominoes. Hence $|B| = 11$. This means that all 11 dominoes lie completely in S_3; that is, the top row is not covered by any dominoes at all. But we can certainly put three more dominoes there, which contradicts the maximality of this arrangement. Hence our assumption was wrong, and one can always place one more domino on a 6×6 chessboard that is covered by 11 dominoes. ∎

Example 4.13. [China 2000, by Jiangang Yao] Let n be a positive integer and define

$$M = \{(x,y) \mid x,y \in \mathbb{N}, 1 \leq x,y \leq n\}.$$

Determine the number of functions f defined on M such that

(i) $f(x,y)$ is a nonnegative integer for any $(x,y) \in M$;

(ii) for $1 \leq x \leq n$, $\sum_{y=1}^{n} f(x,y) = n-1$;

(iii) if $f(x_1,y_1)f(x_2,y_2) > 0$, then $(x_1 - x_2)(y_1 - y_2) \geq 0$.

Solution: There are $\binom{n^2-1}{n-1}$ possible functions f. We treat a function f on M as a matrix $\mathbf{M}_f = (s_{i,j})$ of n rows and n columns such that $f(i,j) = s_{i,j}$. Conditions (i), (ii), (iii) become:

(i') all entries of $\mathbf{M}_f$ are nonnegative integers;

(ii') the sum of the entries in each row of $\mathbf{M}_f$ is $n-1$.

(iii') all of the positive entries of $\mathbf{M}_f$ can be traversed along a path from the upper left entry $s_{1,1}$ to the bottom right entry $s_{n,n}$ by moving only south (down) or east (right) at each step. This is because if $s_{i,j} > 0$ and $s_{i+1,k} > 0$, then $j \leq k$, since $[(i-(i+1)](j-k) \geq 0$. Furthermore, such a path is uniquely determined by $\mathbf{M}_f$ if the path is required to turn south at $s_{i,j}$ if $s_{i,j} > 0$ and $s_{i,k} = 0$ for $j+1 \leq k \leq n$.

We call a matrix satisfying conditions (i'), (ii') and (iii') a *legal matrix*. Clearly, we have a bijection between the set of all functions satisfying the given conditions and the set of legal matrices. It suffices to count the number of legal matrices. With this in mind, we consider the following combinatorial model.

A park is divided into an $n \times n$ grid of unit squares. The park gardener must plant $n-1$ trees in each row of the grid. He can choose not to to plant any trees in a square, and he can choose to plant more than one tree in a square. The gardener works his way from the northwest corner of the park to the southeast corner. He plants one row of trees at a time, and once he finishes a row, he automatically moves south to the next row. The gardener has three kinds of "operation": (1) planting a tree in the square he is in; (2) moving one square east; (3) automatically moving south if he has planted $n-1$ trees in the row he is in (except the last). Furthermore, we do not have to worry about the third kind of operation if the gardener always counts the total number of trees he has planted at any stage, because he moves south if and only if the number has increased by exactly $n-1$. He plants $n-1$ trees in each row for a total of $n(n-1)$ trees planted. Thus, there are $n(n-1)$ operations of the first kind. In order to reach the bottom right corner, he has to make a total number of $n-1$ operations of the second kind. Hence, he has to make a total of $n(n-1)+(n-1) = n^2-1$ operations. Because he can perform these operations in whatever order he chooses, the number of ways in which he can complete his task is $\binom{n^2-1}{n-1}$.

Taking the number of trees planted in each grid square to be the corresponding entry of the matrix $\mathbf{M}_f$, it is not difficult to see that there is a one-to-one correspondence between the legal matrices $\mathbf{M}_f$ and the gardener's tree-planting options.

For example, for $n = 4$, if 1 and e denote the first and the second kind operation, respectively, then the sequences 11ee11111111e11, e11111111111e1e1, and 11111111e11e11e correspond to the pictures of Figure 4.10

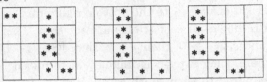

Figure 4.10.

and then to the following three legal matrices of Figure 4.11.

$$
\begin{bmatrix} 2 & 0 & 1 & 0 \\ 0 & 0 & 3 & 0 \\ 0 & 0 & 3 & 0 \\ 0 & 0 & 1 & 2 \end{bmatrix}
\begin{bmatrix} 0 & 3 & 0 & 0 \\ 0 & 3 & 0 & 0 \\ 0 & 3 & 0 & 0 \\ 0 & 1 & 1 & 1 \end{bmatrix}
\begin{bmatrix} 3 & 0 & 0 & 0 \\ 3 & 0 & 0 & 0 \\ 2 & 1 & 0 & 0 \\ 0 & 1 & 2 & 0 \end{bmatrix}
$$

Figure 4.11.

Therefore, the number of matrices $\mathbf{M}_f$, and hence the number of functions f, is equal to $\binom{n^2-1}{n-1}$.   ∎

For positive integers m and n, an ordered m-tuple of nondecreasing positive integers $(a_1, a_2, \ldots, a_m)$ is called an **m-partition** of n if $a_1 + a_2 + \cdots + a_m = n$. The numbers a_i are called the **parts** of the partition; a_m is called the **height** of the partition; and m is called the **length** of the partition. The partition $(a_1, a_2, \ldots, a_m)$ of n is an **increasing** partition if $a_1 < a_2 < \cdots < a_m$. We end this chapter with a result on increasing partitions.

Example 4.14. Let n be a positive integer, and let A denote the set of all increasing partitions of n. Let $a = (a_1, a_2, \ldots, a_m)$ be an element of A. Let $s(a)$ denote the smallest index such that $a_{s(a)}, a_{s(a)+1}, \ldots, a_m$ are consecutive integers; that is, partition a ends with $m - s(a) + 1$ consecutive integers. Further, assume that n cannot be written in the form $\frac{k(3k-1)}{2}$ or $\frac{k(3k+1)}{2}$ for any positive integer k. Let A_1 be a subset of A such that $a \in A_1$ if and only if $a_1 \leq m - s(a) + 1$. Show that $|A| = 2|A_1|$.

Solution: Let A_1' be the subset of A such that $a \in A_1'$ if and only if $a_1 > m - s(a) + 1$. Because either $a_1 \leq m - s(a) + 1$ or $a_1 > m - s(a) + 1$, it is easy to see that A_1 and A_1' is a partition of A. It suffices to show that there is a bijection between A_1 and A_1'. Indeed, we define

$$a = (a_1, a_2, \ldots, a_m) \rightarrow a' = (a_2, \ldots, a_{m-a_1}, a_{m-a_1+1} + 1, \ldots, a_m + 1).$$
$$(*)$$

For example, let $n = 42$ and $a = (3, 5, 7, 8, 9, 10)$. We have $a_1 = 3$, $m = 6$, and $s(a) = 3$, so $a \in A_1$. Then a is mapped to $a' = (5, 7, 9, 10, 11)$. This can be seen more clearly in **Young's diagram**. As shown in Figure 4.12, we remove the three balls in first column (because $a_1 = 3$) and distribute one to each of the three rightmost columns.

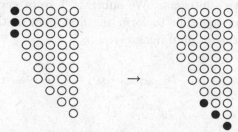

Figure 4.12.

We need to show that (i) $a' \in A_1'$; (ii) the map is injective; and (iii) the map is surjective.

To prove (i), we notice that $m - a_1 + 1 \geq s(a)$, because $a \in A_1$. If $s(a) \geq 2$, then the term a_{m-a_1+1} is well-defined in the mapping $(*)$. If $s(a) = 1$, then n is the sum of m consecutive integers from a_1 to $a_m = a_1 + m - 1$. We must have $a_1 \neq m$ (so $a_1 < m$), because otherwise, we would have $n = m + (m+1) + \cdots + (2m-1) = \frac{m(3m-1)}{2}$, which is not true by the conditions of the problem. (For $k = 3$ and $n = \frac{k(3k-1)}{2}$, Young diagram in Figure 4.13 has no image.)

$\rightarrow$ no image

Figure 4.13.

Hence $a_1 < m$ and $m - a_1 + 1 \geq 2$, and the term a_{m-a_1+1} is again well-defined in the mapping (∗). Thus a' is always well–defined in the map (∗). We note that $s(a') = m - a_1$, and so $m - s(a') = a_1 < a_2$, which is the smallest part of a'. Since a' is clearly in A, we conclude that a' is in A'_1.

To prove (ii), we approach the problem indirectly. Assume to the contrary that there are distinct elements a' and b' in A'_1 such that $a' = b'$. Because the length of the partition is reduced by 1 after the mapping, a and b must have the same length. Let $b = (b_1, b_2, \ldots, b_m)$. Because $m - a_1 = s(a') = s(b') = m - b_1$, $a_1 = b_1$. Thus a and b must have been changed in exactly the same way under the mapping (∗), and a and b must be the same, contradicting our assumption. Hence our assumptions was wrong, the mapping (∗) is indeed injective.

To prove (iii), we show that the mapping (∗) is reversible. Let $a' = (a'_1, a'_2, \ldots, a'_{m-1})$ be an element of A'_1. Then a' ends with $m - s(a')$ consecutive integers. We subtract 1 from each of these parts and put them together to form the first part of the original image of a'. Then the original image is

$$
a = \begin{cases}
(m - s(a'), a'_1, \ldots, a'_{s(a')-1}, a'_{s(a')} - 1, \ldots, a'_{m-1} - 1), \\
\quad \text{if } s(a') > 1; \\
(m - 1, a'_1 - 1, \ldots, a'_{m-1} - 1), \\
\quad \text{if } s(a') = 1.
\end{cases}
$$

It is not difficult to see that the sum of the parts of a is the same as that of a', which is equal to n.

Next we show that a is in A_1. If $s(a') > 1$, then by the definition of $s(a')$, $a'_{s(a')-1}$ and $a_{s(a')}$ are not consecutive integers, so $a'_{s(a')-1} < a'_{s(a')} - 1$. Because a' is in A'_1, $a'_1 > (m-1) - s(a') + 1$. Thus the parts of a are strictly increasing, and it is in A. Write $a = (a_1, a_2, \ldots, a_m)$. To show that a is in A_1, it suffices to show that $a_1 \leq m - s(a) + 1$. Note that a ends with the consecutive integers $a_{s(a')+1} = a'_{s(a')} - 1$, $\ldots$, $a_m = a'_{m-1} - 1$. Consequently, we have $s(a) \leq s(a') + 1$. Thus, $a_1 = m - s(a') \leq m - s(a) + 1$, as desired.

If $s(a') = 1$, then $a'_1 > (m - 1) - s(a') + 1 = m - 1$, because a' is in A'_1. Then $a'_1 \geq m + 1$, because otherwise, $n = m + \cdots + (2m - 2) = \frac{(m-1) \cdot [3(m-1)+1]}{2}$, which is not possible by the conditions of the problem. (For $k = 3$ and $n = \frac{k(3k-1)}{2}$, Young diagram in Figure 4.14 has no original image.)

no original   →

Figure 4.14.

Thus, $m - 1 < a'_1 - 1$, and all of the parts of a are increasing, so a is in A. Write $a = (a_1, a_2, \ldots, a_m)$. Note that a ends with the consecutive integers $a_2 = a'_1 - 1, \ldots, a_m = a'_{m-1} - 1$, so $s(a) \leq 2$. Then $a_1 = m - 1 \leq m - s(a) + 1$, so a is in A_1.

Therefore, for every element a' there is an element a in A_1 with a' as its image. Thus the mapping $(*)$ is a bijection, and $|A| = |A_1| + |A'_1| = 2|A_1|$, as desired. ∎

Exercises 4

4.1. [AIME 2001] A fair die is rolled four times. What is the probability that each of the final three rolls is at least as large as the roll preceding it?

4.2. [St. Petersburg 1989] Tram tickets have six-digit numbers (from 000000 to 999999). A ticket is called *lucky* if the sum of its first three digits is equal to that of its last three digits. A ticket is called *medium* if the sum of all of its digits is 27. Let A and B denote the numbers of lucky tickets and medium tickets, respectively. Find $A - B$.

4.3. Let n be a positive integer. Points $A_1, A_2, \ldots, A_n$ lie on a circle. For $1 \leq i < j \leq n$, we construct $\overline{A_i A_j}$. Let S denote the set of all such segments. Determine the maximum number of intersection points can produced by the elements in S.

4.4. Let $A = \{a_1, a_2, \ldots, a_{100}\}$ and $B = \{1, 2, \ldots, 50\}$. Determine the number of surjective functions f from A to B such that $f(a_1) \leq f(a_2) \leq \cdots \leq f(a_{100})$. What if f does not need to be surjective?

4.5. [AIME 1983] For $\{1, 2, \ldots, n\}$ and each of its nonempty subsets a unique *alternating sum* is defined as follows: Arrange the numbers in the subset in decreasing order and then, beginning with the largest, alternately add and subtract successive numbers. (For example, the alternating sum for $\{1,2,4,6,9\}$ is $9-6+4-2+1 = 6$ and for $\{5\}$, it is simply 5.) Find the sum of all such alternating sums for $n = 7$.

4.6. For a positive integers n denote by $D(n)$ the number of pairs of different adjacent digits in the binary representation of n. For example, $D(3) = D(11_2) = 0$ and $D(21) = D(10101_2) = 4$. For how many positive integers n less than or equal to 2003 is $D(n) = 2$?

4.7. Let

$$\prod_{n=1}^{1996} \left(1 + nx^{3^n}\right) = 1 + a_1 x^{k_1} + a_2 x^{k_2} + \cdots + a_m x^{k_m},$$

where $a_1, a_2, \ldots, a_m$ are nonzero and $k_1 < k_2 < \cdots < k_m$. Find a_{1234}.

4.8. [Putnam 2002] Let n be an integer greater than one, and let T_n be the number of nonempty subsets S of $\{1, 2, 3, \ldots, n\}$ with the property that the average of the elements of S is an integer. Prove that $T_n - n$ is always even.

4.9. Let $p(n)$ denote the number of partitions of n, and let $p(n, m)$ denote the number of partitions of n of length m. Prove that $p(n) = p(2n, n)$.

4.10. A triangular grid is obtained by tiling an equilateral triangle of side length n with n^2 equilateral triangles of side length 1. Determine the number of rhombuses of side length 1 bounded by line segments of the grid.

4.11. Suppose that $P_1 P_2 \ldots P_{325}$ is a regular 325-sided polygon. Determine the number of incongruent triangles of the form $P_i P_j P_k$, where i, j, and k are distinct integers between 1 and 325, inclusive.

4.12. [USAMO 1996, by Kiran Kedlaya] An ordered n-tuple

$$(x_1, x_2, \ldots, x_n)$$

in which each term is either 0 or 1 is called a *binary sequence of length* n. Let a_n be the number of binary sequences of length n containing no three consecutive terms equal to 0, 1, 0 in that order. Let b_n be the number of binary sequences of length n that contain no four consecutive terms equal to 0, 0, 1, 1 or 1, 1, 0, 0 in that order. Prove that $b_{n+1} = 2a_n$ for all positive integers n.

5
Recursions

In this section, we will introduce yet another fundamental technique for solving combinatorial problems.

Suppose we are given a set of objects, S_n whose description involves a parameter n. In order to find the number of elements in S_n, we can view this number as a function of n; that is, we write $|S_n| = f(n)$. We might be able to find an explicit formula for $f(n)$, in terms of n, through relations between $f(n)$ and $f(n-1), \ldots, f(1), f(0)$. Such relations are called **recursive relations** (or **recursions**).

Example 5.1. Given a sphere, a circle is called a great circle if it is the intersection of the sphere with a plane passing through its center. Five distinct great circles dissect the sphere into n regions. Let m and M be the minimum and maximum values of n, respectively. Find m and M. (See Figure 5.1.)

Solution: We consider m and M as functions of k, the number of great circles drawn on the sphere. Let M_k (m_k) denote the maximum (minimum) number of regions in which the sphere can be dissected by k great circles. Then $M_1 = m_1 = 2$.

Assume that there are k great circles drawn on the sphere, and the sphere is dissected into a_k regions. When the $(k+1)^{\text{st}}$ great circle is drawn, it will cut through a certain number of regions. One more region will be produced by each region it cuts through. The $(k+1)^{\text{st}}$ great circle intersects the other k great circles, and every pair of adjacent intersection points forms an arc that cuts through an existing region.

Figure 5.1.

Note that two distinct great circles must intersect at two points. Hence any number of great circles have a minimum of two intersection points, and $m_{k+1} \geq m_k + 2$. Thus, $m_5 \geq m_4 + 2 \geq m_3 + 2 + 2 \geq \cdots \geq m_1 + 8 = 10$. It is easy to see that we can obtain 10 regions by drawing five great circles through two diametrically opposite points, so $m = m_5 = 10$.

On the other hand, the $(k+1)^{\text{st}}$ great circle will have at most $2k$ intersection points with the first k great circles, leading to $2k$ new regions. By rotating the circle slightly along any axis, we can indeed obtain $2k$ new intersections. Hence $M_{k+1} = M_k + 2k$, and we obtain $M = M_5 = M_1 + 2(1 + 2 + 3 + 4) = 22$.   ∎

Under most circumstances, it is necessary to build some simple bijections in order to set up recursive relations.

Example 5.2. [MOSP 1999, by Cecil Rousseau] A finite set of positive integers is *fat* if each of its members is at least as large as the number of elements in the set. (The empty set is considered to be fat.) Let a_n denote the number of fat subsets of $\{1, 2, \ldots, n\}$ that contain no two consecutive integers, and let b_n denote the number of subsets of $\{1, 2, \ldots, n\}$ in which any two elements differ by at least three. Prove that $a_n = b_n$ for all $n \geq 0$.

Solution: Let $S_n = \{1, 2, \ldots, n\}$. Let A_n denote the set of all fat subsets of S_n that contain no two consecutive integers. Let B_n denote the set of all subsets of S_n in which any two elements differ by at least three. Then $a_n = |A_n|$ and $b_n = |B_n|$. Let C_n be the set of those fat subsets in A_n that contain n, and let $c_n = |C_n|$.

Let s be an element of B_n. If n is not in s, then s is an element of B_{n-1}. If n is in s, then $n-1$ and $n-2$ are not in s, and the set $s - \{n\}$ is an element of B_{n-3}. Thus, $b_n = b_{n-1} + b_{n-3}$.

Now let s be an element of A_n. If n is not in s, then s is an element of A_{n-1}. If n is in s, then s is in C_n. Hence $a_n = a_{n-1} + c_n$. We will show that $c_n = a_{n-3}$ by defining a bijection between C_n and A_{n-3}.

For an element c in C_n, we map c to a as

$$c = \{n_1 < n_2 < \cdots < n_i < n\} \to a = \{n_1 - 1 < n_2 - 1 < \cdots < n_i - 1\}$$

for $i \geq 1$ and $c = \{n\} \to a = \emptyset$. From the fact that $n_i \leq n - 2$ we have $n_i - 1 \leq n - 3$. From $n_1 \geq i + 1$ (because c has $i + 1$ elements) we know that $n_1 - 1 \geq i$. Hence $a \in A_{n-3}$. It is then easy to check that this is indeed a bijection. The result follows by strong induction. ∎

We will present two approaches to each of the next two examples. The first solution to Example 5.3 was provided by David Vincent while he was taking a combinatorics course at PEA. The first solution to Example 5.4 was given by Reid Barton when he was helping the authors to edit [7]. Both solutions are clever. On the other hand, the second solutions to both examples apply recursive relations of three sequences with simple relations between them. The reader may start appreciating the power of recursive relations after a comparison of the two methods.

Example 5.3.   Using the digits 0, 1, 2, 3, and 4, how many ten-digit sequences can be written so that the difference between any two consecutive digits is 1?

First Solution: We call an n-digit sequence *good* if it uses the digits 0, 1, 2, 3, 4 in such a way that the difference between any two consecutive digits is 1. Let $a_1 a_2 \ldots a_{10}$ be a good ten-digit sequence. The numbers in the sequence must alternate in parity. Hence either $a_1 a_3 a_5 a_7 a_9$ or $a_2 a_4 a_6 a_8 a_{10}$ contain all odd numbers. Let $b_1 b_2 b_3 b_4 b_5$ denote the odd subsequence. There are two possible values (1 or 3) for b_1. Now we consider the good nine-digit sequence

$$b_1 c_1 b_2 c_2 b_3 c_3 b_4 c_4 b_5$$

generated from $b_1 b_2 b_3 b_4 b_5$. Then $b_i = b_{i+1}$ if and only if there are two possible values ($b_i \pm 1$) for c_i; and $b_i \neq b_{i+1}$ if and only if there is one possible value ($(b_i + b_{i+1})/2$) for c_i. Given the value of b_i, $i = 1, \ldots, 4$, a good sequence can be extended in exactly three ways, by setting $c_i = b_i + 1$, $b_{i+1} = b_i$, or $c_i = b_i - 1$, $b_{i+1} = b_i$, or $c_i = 2$, $b_{i+1} = 4 - b_i$. Hence there are 3^4 ways to build a good sequence once the value of b_1 is chosen.

For each good nine-digit sequence, we can construct four distinct good 10-digit sequences, since there are two possible places for the remaining digit and two possible values for the remaining digit (that is why we chose the odd subsequence in the beginning). Indeed, the final sequence has one of the following forms

$$(b_1 \pm 1)b_1c_1b_2c_2b_3c_3b_4c_4b_5 \quad \text{or} \quad b_1c_1b_2c_2b_3c_3b_4c_4b_5(b_5 \pm 1).$$

Hence the answer is $2 \cdot 3^4 \cdot 4 = 648$.   ∎

Second Solution: Let A_n denote the number of good n-digit sequences that end in either 0 or 4; let B_n denote the number of good n-digit sequences that end in either 1 or 3; and let C_n denote the number of good n-digit sequences that end in 2. Then for nonnegative integers n:

(a) $A_{n+1} = B_n$, because each sequence in A_{n+1} can be converted into a sequence in B_n by deleting its last digit.

(b) $B_{n+1} = A_n + 2C_n$, because each sequence in A_n can be converted into a sequence in B_{n+1} by adding a 1 (if it ends with 0) or a 3 (if it ends with 4) to its end, and each sequence in C_n can be converted into a sequence in B_{n+1} by adding a 1 or a 3 to it.

(c) $C_{n+1} = B_n$, because each sequence in C_{n+1} can be converted into a sequence in B_n by deleting the 2 at its end.

Hence $B_{n+1} = 3B_{n-1}$ for $n \geq 2$. It is easy to see that $B_1 = 2$ and $B_2 = 4$. Hence $B_{2n+1} = 2 \cdot 3^n$ and $B_{2n} = 4 \cdot 3^{n-1}$. It follows that there are

$$A_{10} + B_{10} + C_{10} = 2B_9 + B_{10} = 4 \cdot 3^4 + 4 \cdot 3^4 = 648$$

good ten-digit sequences.   ∎

Example 5.4.   [Romania 1998] Let A_n denote the set of codes of length n, formed by using the letters a, b, and c, none of which contains consecutive appearances of a's or consecutive appearances of b's. Let B_n denote the set of codes of length n formed by using the letters a, b, and c, none of which contains three consecutive letters that are distinct (so at least two of the three letters are the same). Prove that $|B_{n+1}| = 3|A_n|$ for all $n \geq 1$.

First Solution: Replace a, b, and c by 1, 2, and 0, respectively, and codes of length n by ordered n-tuples. Let S_n be the set of all ordered n-tuples in which each term is 0, 1, or 2. Consider the function $\Delta : S_{n+1} \to S_n$ defined as follows:

$$\Delta(x_1, \ldots, x_{n+1}) = (y_1, \ldots, y_n),$$

where $y_i \equiv x_{i+1} - x_i$ (mod 3). Observe that x_{i-1}, x_i, x_{i+1} are distinct if and only if $0 \not\equiv x_i - x_{i-1} \not\equiv x_{i+1} - x_i \not\equiv 0$ (mod 3), i.e., $\{x_i - x_{i-1}, x_{i+1} - x_i\} \equiv \{1.2\}$ (mod 3). Thus $(x_1, \ldots, x_{n+1})$ contains no three consecutive distinct elements if and only if $\Delta(x_1, \ldots, x_{n+1})$ contains no two consecutive 1's or 2's. Therefore, $\Delta(B_{n+1}) = A_n$. A sequence x in S_{n+1} such that $\Delta(x) = y$ is uniquely determined by y and the first symbol in x, which can take any of the 3 possible values. Thus we conclude that $|\Delta^{-1}(\{y\})| = 3$ for all $y \in S_n$, and we have $|B_{n+1}| = 3|A_n|$.

Second Solution: Let X_n, Y_n, and Z_n be the subsets of A_n containing codes that end with the letters a, b, and c, respectively. For each positive integer n, we have the following relations:

(a) $|A_n| = |X_n| + |Y_n| + |Z_n|$, because X_n, Y_n, Z_n is a partition of A_n;

(b) $|X_{n+1}| = |Y_n| + |Z_n|$, because each code in X_{n+1} can be converted into a code in either Y_n or Z_n by deleting its last letter a;

(c) $|Y_{n+1}| = |Z_n| + |X_n|$, because each code in Y_{n+1} can be converted into a code in either Z_n or X_n by deleting its last letter b;

(d) $|Z_{n+1}| = |X_n| + |Y_n| + |Z_n|$, because each code in Y_{n+1} can be converted into a code in either X_n, Y_n, or Z_n by deleting its last letter c.

By (a) and (d), we have $|Z_{n+1}| = |A_n|$. Therefore,

$$|A_{n+2}| = |X_{n+2}| + |Y_{n+2}| + |Z_{n+2}|$$
$$= 2|X_{n+1}| + 2|Y_{n+1}| + 3|Z_{n+1}| = 2|A_{n+1}| + |A_n|,$$

with $|A_1| = 3$ and $|A_2| = 7$ (because $A_2 = \{ab, bc, ca, ba, ac, cb, cc\}$).

Let T_n be the set of all codes in B_n that end with letters aa, bb, or cc. Let $U_n = B_n - T_n$, i.e., U_n contains those codes in T_n that end in ab, ac, ba, bc, ca, or cb. For each positive integer n, we have the following relations:

(a') $|B_n| = |T_n| + |U_n|$, because T_n and U_n partition B_n;

(b') $|B_{n+1}| = 3|T_n| + 2|U_n|$, because each code in T_n can be converted into a code in B_{n+1} by adding either an a, or b, or c at the end, and each code in U_n can be converted into a code in B_{n+1} by adding one of its last two letters to it.

(c$'$) $|T_{n+1}| = |B_n|$, because each code in B_n can be converted into a code in T_{n+1} by adding its last letter to it.

Consequently,

$$|B_n| - |B_{n-1}| = |B_n| - |T_n| = |U_n|$$
$$= 3(|T_n| + |U_n|) - (3|T_n| + 2|U_n|) = 3|B_n| - |B_{n+1}|,$$

from which we get $|B_{n+1}| = 2|B_n| + |B_{n-1}|$. Any two-letter code is an element of B_2, so $|B_2| = 3^2 = 9 = 3|A_1|$. A three-letter code is in B_3 if and only if it is not a permutation of $\{a, b, c\}$, so $|B_3| = 3^3 - 3! = 21 = 3|A_2|$. Thus $|A_n|$ and $|B_{n+1}|$ satisfy the same linear recursive relation, with the initial terms of $|B_{n+1}|$ being three times al large as the corresponding initial terms of $|A_n|$. Thus $|B_{n+1}| = 3|A_n|$.   ∎

There are two practical difficulties in applying recursive relations. First, it is not always easy to find recursive relations. Second, it is hard to transform the recursive relations that a sequence satisfies into explicit formulas to describe the sequence. If we overcome the first difficulty and the term we are looking for does not come too late in the sequence, we can overcome the second difficulty by brute force: direct calculation!

Example 5.5. [AIME 1995] What is the probability that in the process of repeatedly flipping a fair coin, one will encounter a run of 5 heads before one encounters a run of 2 tails?

Solution: A *successful string* is a sequence of H's and T's in which $HHHHH$ appears before TT does. Each successful string must belong to one of the following three types:

(a) those that begin with T, followed by a successful string that begins with H;

(b) those that begin with H, HH, HHH, or $HHHH$, followed by a successfully string that begins with T.

(c) the string $HHHHH$.

Let p_h denote the probability of obtaining a successful string that begins with H, and let p_t denote the probability of obtaining a successful string that begins with T. The three types of winning strings allow us to build recursive relations in the backward direction. More precisely, a winning string of type (a) is of the form of $TH \ldots$, which can be mapped one-to-one to a winning string of the form $H \ldots$. Thus,

$$p_t = \frac{1}{2}p_h.$$

On the other hand, a winning string beginning with k $(1 \le k \le 4)$ consecutive H's can also mapped one-to-one to a winning string beginning with T. It follows that

$$p_h = \left(\frac{1}{2} + \frac{1}{4} + \frac{1}{8} + \frac{1}{16}\right)p_t + \frac{1}{32}.$$

Solving the last two equations simultaneously, we find that

$$p_h = \frac{1}{17} \quad \text{and} \quad p_t = \frac{1}{34}.$$

Thus the probability of obtaining five heads before obtaining two tails is $p_h + p_t = \frac{3}{34}$.   ∎

There are a few recursive relations that can be dealt with mechanically.

Example 5.6. [AIME 1985] Let A, B, C, and D be the vertices of a regular tetrahedron, each of whose edges measures 1 meter (Figure 5.2). A bug, starting from vertex A, observes the following rule: At each vertex it chooses one of the three edges meeting at that vertex, each edge being equally likely, and crawls along that edge to the vertex at its opposite end. Find the probability that the bug is at vertex A when it has crawled exactly 7 meters.

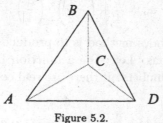

Figure 5.2.

Solution: For $n = 0, 1, 2, \ldots$, let a_n be the probability that the bug is at vertex A after crawling exactly n meters. Then we have the recursive relation

$$a_{n+1} = \frac{1}{3}(1 - a_n), \tag{†}$$

because the bug can be at vertex A after crawling $n+1$ meters if and only if

(a) it was not at A following a crawl of n meters (this has probability $1 - a_n$) and

(b) from one of the other vertices it heads toward A (this has probability $\frac{1}{3}$).

Now, since $a_0 = 1$ (i.e., the bug starts at vertex A), from the recursive relation (†), we have $a_1 = 0$, $a_2 = \frac{1}{3}$, $a_3 = \frac{2}{9}$, $a_4 = \frac{7}{27}$, $a_5 = \frac{20}{81}$, $a_6 = \frac{61}{243}$, and the desired probability is $p = a_7 = \frac{182}{729}$. ∎

We notice that relation (†) is close to the recursive relation of a geometric progression with common ratio $-\frac{1}{3}$. Let x be a real number such that

$$a_{n+1} - x = -\frac{1}{3}(a_n - x) = \frac{1}{3}(x - a_n).$$

After substitution in the relation (†), we obtain

$$x = \frac{1}{3}(1 - x),$$

implying that $x = \frac{1}{4}$. Consequently, we may write (†) in the form

$$a_n - \frac{1}{4} = -\frac{1}{3}\left(a_{n-1} - \frac{1}{4}\right).$$

Hence $b_n = a_n - \frac{1}{4}$ is a geometric progression with common ratio $-\frac{1}{3}$ and $b_0 = a_0 - \frac{1}{4} = \frac{3}{4}$, implying that

$$a_n = \frac{1}{4} + \left(-\frac{1}{3}\right)^n \left(\frac{3}{4}\right).$$

The heart of the above method is to produce a geometric progression through recursions. Let f be a function from the nonnegative integers to the real numbers. If there are real constants $a_1, a_2, \ldots, a_k$ such that

$$f(n) - a_1 f(n-1) - a_2 f(n-2) - \cdots - a_k f(n-k) = 0 \tag{$*$}$$

for all integers $n \geq k$, this relation is called a *constant homogeneous recursion of degree k*. If there are real constants $a_1, a_2, \ldots, a_k$ such that

$$f(n) - a_1 f(n-1) - a_2 f(n-2) - \cdots - a_k f(n-k) = g(n) \quad (*')$$

for all integers $n \geq k$, this relation is called a *constant inhomogeneous recursion of degree k*. The equation

$$x^k - a_1 x^{k-1} - a_2 x^{k-2} - \cdots - a_k = 0 \quad (**)$$

is called the *characteristic equation* of the recursion $(*)$. The roots of equation $(**)$ are called the *characteristic roots* of the recursion $(*)$. We have the following theorems.

Theorem 5.1. *Let $z \neq 0$ be a complex number. Then $f(n) = z^n$ satisfies $(*)$ if and only if z satisfies $(**)$; that is, z is a characteristic root of $(*)$.*

Proof: Note that $f(n) = z^n$ satisfies $(*)$ if and only if

$$z^{n-k}(z^k - a_1 z^{k-1} - a_2 z^{k-2} - \cdots - a_k) = 0,$$

which is equivalent to z satisfying $(**)$, because $z \neq 0$. ∎

Because recursion $(*)$ is linear, we easily have the following result.

Theorem 5.2. *If both $f_1(n)$ and $f_2(n)$ satisfy recursion $(*)$, then so do all of their linear combinations; that is, $c_1 f_1(n) + c_2 f_2(n)$ satisfies $(*)$ for any constants c_1 and c_2.*

Theorem 5.3. *If recursion $(*)$ has k distinct characteristic roots $z_1, z_2, \ldots, z_k$, then all of the functions $f(n)$ satisfying $(*)$ are linear combinations of $z_1^n, z_2^n, \ldots, z_k^n$, that is,*

$$f(n) = c_1 z_1^n + c_2 z_2^n + \cdots + c_k z_k^n,$$

where $c_1, c_2, \ldots, c_k$ are real constants.

Proof: By Theorems 5.1 and 5.2, we know that the linear combinations of $z_1^n, z_2^n, \ldots, z_k^n$ satisfy $(*)$.

Assume that $f(n)$ satisfies $(*)$. Then $f(n)$ is determined by the initial values $f(0), f(1), \ldots, f(k-1)$. Consider the linear system

$$\begin{bmatrix} 1 & 1 & \cdots & 1 \\ z_1 & z_2 & \cdots & z_k \\ z_1^2 & z_2^2 & \cdots & z_k^2 \\ \cdots & \cdots & & \cdots \\ z_1^{k-1} & z_2^{k-1} & \cdots & z_k^{k-1} \end{bmatrix} \cdot \begin{bmatrix} c_1 \\ c_2 \\ c_3 \\ \cdots \\ c_k \end{bmatrix} = \begin{bmatrix} f(0) \\ f(1) \\ f(2) \\ \cdots \\ f(k-1) \end{bmatrix}.$$

This system is solvable if and only if the determinant of the first matrix is nonzero. But the first matrix is the famous **Vandermonde matrix**, whose determinant is equal to

$$\prod_{1 \le i < j \le k} (z_j - z_i).$$

This value is nonzero if and only if all of the characteristic roots $z_1, z_2, \ldots, z_k$ are distinct. Hence we can find real numbers $c_1, c_2, \ldots, c_k$ such that

$$f(n) = c_1 z_1^n + c_2 z_2^n + \cdots + c_k z_k^n,$$

for $n = 0, 1, \ldots, k - 1$. Thus the function $c_1 z_1^n + c_2 z_2^n + \cdots + c_k z_k^n$ satisfies the given recursion and agrees with $f(n)$ on the k initial values $n = 0, 1, \ldots, k - 1$, which implies that

$$f(n) = c_1 z_1^n + c_2 z_2^n + \cdots + c_k z_k^n,$$

for all $n \ge 0$.   ∎

Theorem 5.4. *If recursion* (∗) *has* m *distinct characteristic roots* $z_1, z_2, \ldots, z_m$, *such that the multiplicity of the root* z_i *in* (∗∗) *is* e_i, *for* $i = 1, \ldots, k$, *then all functions* $f(n)$ *satisfying* (∗) *are linear combinations of*

$$z_1^n, n z_1^n, \ldots, n^{e_i - 1} z_1^n;$$
$$z_2^n, n z_2^n, \ldots, n^{e_2 - 1} z_2^n;$$
$$\cdots\cdots\cdots;$$
$$z_m^n, n z_m{}^n, \ldots, n^{e_m - 1} z_m{}^n.$$

This theorem can be proved using a generalized **Vandermonde matrix** and a bit more knowledge about polynomials. Because they are not the main focus of this book, we will omit the proof of Theorem 5.4.

Because the recursion (∗′) is also linear, we easily have the following Theorem.

Theorem 5.5. *A function* $f(n)$ *satisfying a constant inhomogeneous recursion* (∗′) *can be written in the form*

$$f(n) = f_1(n) + f_2(n),$$

where $f_1(n)$ satisfies a constant homogeneous recursion (∗), and $f_2(n)$ is a particular function satisfying (∗′).

While we have shown systematic ways to find all solutions of a given constant homogeneous recursion (∗), it is also interesting to find particular solutions of the constant inhomogeneous recursion (∗′). If $g(n)$ is a polynomial in n of degree d, we should try to find a particular solution that is a polynomial in n of degree at most d. We can usually determine the coefficients of such a solution from $d+1$ initial conditions. If $g(n)$ is an exponential function in n with its base not equal to any of the characteristic roots, then we should try to find a particular solution that is exponential function in n in the same base. Because it is not our intention to turn this book into a list of rules, we are not going to list all of these special cases. Instead, we will present more examples for the reader to enjoy.

Example 5.7. [IMO 1979] Let A and E be opposite vertices of a regular octagon (Figure 5.3). A frog starts jumping at vertex A. From any vertex of the octagon except E, it may jump to either of the two adjacent vertices. When it reaches vertex E, the frog stops and stays there. Let u_n be the number of distinct paths of exactly n jumps ending at E. Prove that $u_{2n-1} = 0$,

$$u_{2n} = \frac{1}{\sqrt{2}}(x^{n-1} - y^{n-1}), \quad n = 1, 2, 3, \ldots,$$

where $x = 2 + \sqrt{2}$ and $y = 2 - \sqrt{2}$. (Note that a path of n jumps is a sequence of vertices $(P_0, \ldots, P_n)$ such that

(i) $P_0 = A, P_n = E$;

(ii) for every $i, 0 \le i \le n-1, P_i$ is distinct from E; and

(iii) for every $i, 0 \le i \le n-1, P_i$ and P_{i+1} are adjacent.

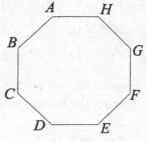

Figure 5.3.

Solution: Let $a_n, b_n, c_n, d_n, e_n, f_n, g_n$, and h_n be the number of distinct paths of exactly n jumps ending at A, B, C, D, E, F, G, and H, respectively. Then $e_n = u_n$. By symmetry, we know that $b_n = h_n$, $c_n = g_n$, and $d_n = f_n$. It is also not difficult to see that

$$e_n = d_{n-1} + f_{n-1} = 2d_{n-1},$$

$$d_n = c_{n-1},$$

$$a_n = 2b_{n-1},$$

$$c_n = b_{n-1} + d_{n-1},$$

$$b_n = a_{n-1} + c_{n-1}.$$

From the first two equations, we obtain $d_{n-1} = \frac{1}{2}e_n$ and $c_{n-1} = d_n = \frac{1}{2}e_{n+1}$. Substituting these two relations into the fourth equation gives $b_{n-1} = \frac{1}{2}(e_{n+2} - e_n)$. From the third equation, we obtain $a_n = e_{n+2} - e_n$. Then the fifth equation reads

$$e_{n+3} - 4e_{n+1} + 2e_{n-1} = 0.$$

Thus we may consider the characteristic equation

$$x^4 - 4x^2 + 2 = 0.$$

The four roots are $\pm\sqrt{2 \pm \sqrt{2}}$. By Theorem 5.3,

$$u_n = e_n = as^n + b(-s)^n + ct^n + d(-t)^n,$$

where $s = \sqrt{x} = \sqrt{2 + \sqrt{2}}$ and $t = \sqrt{y} = \sqrt{2 - \sqrt{2}}$. Note that $e_1 = e_2 = e_3 = 0$ and $e_4 = 2$. Hence

$$(a - b)s + (c - d)t = 0,$$

$$(a + b)s^2 + (c + d)t^2 = 0,$$

$$(a - b)s^3 + (c - d)t^3 = 0,$$

$$(a + b)s^4 + (c + d)t^4 = 2.$$

From the first and third equations, we obtain $a = b$ and $c = d$. This implies that $u_{2n-1} = 0$. Solving the second and fourth equations gives

$$a = b = \frac{t^2}{4\sqrt{2}} \quad \text{and} \quad c = d = -\frac{s^2}{4\sqrt{2}}.$$

Consequently,

$$u_{2n} = 2as^{2n} + 2ct^{2n} = \frac{s^2t^2}{2\sqrt{2}}(s^{2n-2} - t^{2n-2}) = \frac{1}{\sqrt{2}}(x^{n-1} - y^{n-1}),$$

as desired.   ∎

We can also write these recursions in matrix form:

$$\begin{bmatrix} a_n \\ b_n \\ c_n \\ d_n \\ e_n \end{bmatrix} = \mathbf{M} \cdot \begin{bmatrix} a_{n-1} \\ b_{n-1} \\ c_{n-1} \\ d_{n-1} \\ e_{n-1} \end{bmatrix} = \mathbf{M}^{n-1} \cdot \begin{bmatrix} 1 \\ 0 \\ 0 \\ 0 \\ 0 \end{bmatrix},$$

where

$$\mathbf{M} = \begin{bmatrix} 0 & 2 & 0 & 0 & 0 \\ 1 & 0 & 1 & 0 & 0 \\ 0 & 1 & 0 & 1 & 0 \\ 0 & 0 & 1 & 0 & 0 \\ 0 & 0 & 0 & 2 & 0 \end{bmatrix}.$$

Readers familiar with linear algebra might want to investigate the relations between recursions, eigenvalues, and Jordan matrices.

Example 5.8. [USAMO 1990] Let n be a positive integer. Find the number of positive integers whose base n representation consists of distinct digits with the property that except for the leftmost digit, every digit differs by ± 1 from some digit further to the left.

First Solution: We apply a double recursion. Let a_n denote the number of suitable base n representations. Now we consider the suitable base $(n + 1)$ representations. These representations can be put into two disjoint classes:

(1) digit n is not in the representation, so this representation is also a suitable representation in base n; and

(2) digit n is in the representation.

Assume that there are b_{n+1} suitable representations in the second class. Then $a_{n+1} = a_n + b_{n+1}$. If a representation in the second class has k digits, those digits must be $n, n-1, \ldots, n-k+1$. We leave it as an exercise to show that $b_{n+1} = 2^{n+1} - 2$. Thus,

$$a_{n+1} = a_n + 2^{n+1} - 2.$$

Hence $a_{n+1} = a_1 + (2^{n+1} + 2^n + \cdots + 2^2) - 2n$. Because $a_1 = 0$, it follows that $a_{n+1} = 2^{n+2} - 4 - 2n$ or

$$a_n = 2^{n+1} - 4 - 2(n-1) = 2^{n+1} - 2n - 2.$$

■

Second Solution: A clever solution to this problem applies a bijection. Define $F(n)$ as the number of suitable base n integers plus the number of base n digit strings that begin by 0 but are otherwise suitable. One can easily find that $F(1) = 1$, $F(2) = 4$, and $F(3) = 11$. To establish a recursion, note that the suitable base $(n+1)$ digit strings fall into three disjoint classes:

(a) a single digit $0, 1, \ldots, n$;

(b) a suitable digit string in base n, followed by the next largest unused digit; and

(c) a suitable digit string in base n, with each digit increased by 1, followed by the next-smallest unused digit.

We leave it to the reader to check the details of the above arguments. It follows that

$$F(n+1) = n + 1 + 2F(n).$$

We rewrite this recursion as

$$F(n+1) + a(n+1) + b = 2(F(n) + an + b)$$

for some real numbers a and b, implying that $an + b - a = n + 1$, so $a = 1$ and $b = 2$. Hence $G(n) = F(n) + n + 2$ is a geometric progression with common ratio 2, and $G(1) = 4$. Consequently, $G(n) = 2^{n+1}$ and $F(n) = 2^{n+1} - n - 2$. The only suitable base n strings that are not suitable integers in the problem are $0, 01, 012, \ldots, 012\ldots(n-1)$, so the answer is $a_n = 2^{n+1} - 2n - 2$.   ■

We can also use the procedure described in Theorem 5.5 to find an explicit formula for $F(n)$. Note that the general solution of the homogeneous recursion $F(n+1) = 2F(n)$ is $c \cdot 2^n$. Setting $F(n) = an + b$ in $F(n+1) = n + 1 + 2F(n)$ yields $(a, b) = (-1, -2)$. Hence the general solution of $F(n+1) = n + 1 + 2F(n)$ is $F(n) = c \cdot 2^n - n - 2$, where c can be determined from the initial values of $F(n)$.

Example 5.9.   [China 2001, by Chengzhang Li] Let $P_1 P_2 \ldots P_{24}$ be a regular 24-sided polygon inscribed in a circle ω with circumference

24. Determine the number of ways in which we can choose sets of eight distinct vertices $\{P_{i_1}, P_{i_2}, \ldots, P_{i_8}\}$ such that none of the arcs $P_{i_j} P_{i_k}$ has length 3 or 8.

Solution: We set up the following array of vertices:

$$
\begin{array}{cccccccc}
P_1 & P_4 & P_7 & P_{10} & P_{13} & P_{16} & P_{19} & P_{22} \\
P_9 & P_{12} & P_{15} & P_{18} & P_{21} & P_{24} & P_3 & P_6 \\
P_{17} & P_{20} & P_{23} & P_2 & P_5 & P_8 & P_{11} & P_{14}
\end{array}
$$

We color the vertices in the first row red, the vertices in the second row green, and the vertices in the third row blue. It is not difficult to see that exactly one vertex should be picked from each column, and that vertices picked from the ith and $(i+1)^{\text{th}}$ columns for $1 \le i \le 8$ (where the 9th column is the first column) must be different colors. Hence the problem is equivalent to the following problem:

Let $A_1 A_2 \ldots A_n$ $(n \ge 3)$ be a regular n-sided polygon with O as its center. Triangular regions $OA_i A_{i+1}$, $1 \le i \le n$ $(A_{n+1} = A_1)$, are to be colored in one of k $(k \ge 3)$ colors in such a way that adjacent regions are colored differently. Let $p_{n,k}$ denote the number of such colorings. Find $p_{8,3}$.

There are k ways to color region $OA_1 A_2$, then $k-1$ ways to color regions $OA_2 A_3$, $OA_3 A_4$, and so on. We have to be careful about coloring region $OA_n A_1$, because it is possible that it is the same color as region $OA_1 A_2$. But then we simply end up with a valid coloring for $n-1$ regions by viewing region $OA_n A_1 A_2$ as one region. There is a clear bijection between this special kind of invalid coloring of n regions and valid colorings of $n-1$ regions. Hence $p_{n,k} = k(k-1)^{n-1} - p_{n-1,k}$. Note that $p_{3,k} = k(k-1)(k-2)$. It follows that

$$
\begin{aligned}
p_{n,k} &= k(k-1)^{n-1} - k(k-1)^{n-2} + k(k-1)^{n-3} - \cdots \\
&\quad + (-1)^{n-4} k(k-1)^3 + (-1)^{n-3} k(k-1)(k-2) \\
&= k \cdot \frac{(k-1)^n + (-1)^{n-4}(k-1)^3}{1+(k-1)} + (-1)^{n-3} k(k-1)(k-2) \\
&= (k-1)^n + (-1)^n (k-1)^3 + (-1)^{n-1} k(k-1)(k-2) \\
&= (k-1)^n + (-1)^n (k-1)[(k-1)^2 - k(k-2)] \\
&= (k-1)^n + (-1)^n (k-1).
\end{aligned}
$$

Hence $p_{8,3} = 2^8 + 2 = 258$ is the desired answer. ∎

Example 5.10. [MOSP 2001, from Kvant] Prove that there are more than 8^n n-digit numbers not containing any sequence of digits (of any length) twice in a row.

Solution: We call the numbers satisfying the condition of the problem *good*. Let $f(n)$ denote the number of good numbers. We will show that $f(n) \geq 8f(n-1)$ by using induction on n. Then, since $f(1) = 9 > 8$, we will have $f(n) \geq 8^{n-1}f(1) > 8^n$, as desired.

The base case $n = 2$ is trivial, since $f(1) = 9$ and $f(2) = 90-9 = 81$.

Assume that $f(n) \geq 8f(n-1)$ for $n = 2, 3, \ldots, k-1$ for some integer $k \geq 3$. We need to show that $f(k) \geq 8f(k-1)$. Consider a k-digit good number. Its first $k-1$ digits must form a good number. Thus, we may add any of the 10 possible digits to the right of an arbitrary $(k-1)$-digit good number in order to get a k-digit good number, unless we create a repeated string of the form

$$\underbrace{* * \cdots *}_{k-2r \text{ digits}} \underbrace{a_1 a_2 \ldots a_r}_{r \text{ digits}} \underbrace{a_1 a_2 \ldots a_{r-1}}_{r-1 \text{ digits}} a_r.$$

How often can we have a repeated string of length r? The answer is 0 for $r > \lfloor \frac{k}{2} \rfloor$ and $f(k-r)$ for $1 \leq r \leq \lfloor \frac{k}{2} \rfloor$. This is because there is a bijection between the set of good $(k-r)$-digit numbers and the set of bad k-digit numbers that have exactly two repeated strings of length r at the end. Therefore,

$$f(k) = 10f(k-1) - \sum_{r=1}^{\lfloor \frac{k}{2} \rfloor} f(k-r) \geq 10f(k-1) - \sum_{r=1}^{k-1} f(k-r)$$

$$= 10f(k-1) - (f(k-1) + f(k-2) + \cdots + f(1))$$

$$\geq 10f(k-1) - \left(f(k-1) + \frac{1}{8}f(k-1) + \frac{1}{8^2}f(k-1) + \cdots \right)$$

$$> 10f(k-1) - \left(f(k-1) + \frac{1}{2}f(k-1) + \frac{1}{2^2}f(k-1) + \cdots \right)$$

$$= 10f(k-1) - 2f(k-1) = 8f(k-1),$$

as desired.   ∎

The next example involves a two-variable recursion.

Example 5.11. [China 2002] Let n be a positive integer. Express

$$\sum (a_1 a_2 \cdots a_n)$$

in closed form, where the sum is taken over all n-term sequences of positive integers such that $a_1 = 1$ and

$$a_{i+1} \leq a_i + 1$$

for all $1 \leq i \leq n - 1$.

Solution: Let

$$f(m, n) = \sum (a_1 a_2 \cdots a_n),$$

where the sum is taken over all n-term sequences of positive integers such that $a_1 = m$ and

$$a_{i+1} \leq a_i + 1 \quad \text{for all} \quad 1 \leq i \leq n - 1.$$

We want to find $f(1, n)$. If $a_1 = m$, then $1 \leq a_2 \leq m + 1$. Hence we have the recursion

$$\begin{aligned}
f(m, n + 1) &= m \sum (a_2 a_3 \cdots a_{n+1}) \\
&= m[f(1, n) + f(2, n) + \cdots + f(m + 1, n)].
\end{aligned}$$

In order to find $f(m, n)$ in closed form, it seems that $f(m, n+1)$ and the partial sum $f(1, n) + f(2, n) + \cdots + f(m + 1, n)$ can be written in the same form. Theorem 3.2 (e) and (g) comes to mind. Writing

$$(2n - 1)!! = \prod_{i=1}^{n} (2i - 1),$$

we will prove that

$$f(m, n) = \binom{m + 2n - 2}{2n - 1} \cdot (2n - 1)!!$$

by using induction on n.

The base case is trivial because $f(m, 1) = m$. Now assume that $f(m, n)$ satisfies the above identity for positive integers $1 \leq n \leq k$.

We have

$$f(m, k+1)$$
$$= m[f(1, k) + f(2, k) + \cdots + f(m+1, k)]$$
$$= m \cdot (2k-1)!! \cdot \left[\binom{2k-1}{2k-1} + \binom{2k}{2k-1} + \cdots + \binom{m+2k-1}{2k-1} \right]$$
$$= m \cdot (2k-1)!! \cdot \binom{2k+m}{2k}$$
$$= (2k+1)!! \binom{2k+m}{2k+1},$$

by Theorem 3.2 (g) and (e). Thus our induction is complete. Hence, the answer is $f(1, n) = (2n-1)!!$.   ∎

Example 5.12. Let n be a positive integer. We are given an $n \times n$ chessboard and n rooks. Let A and B be two points on the edges of the squares determining the chessboard. A *path connecting A and B* is a continuous curve along the edges of the squares of the chessboard with A and B as endpoints. The *length* of the curve is the length of the path. Claudia is asked to complete the following task. First, she places all of the rooks on the chessboard so that the rooks cannot attack each other. Next, she draws a path of length $2n$ connecting the top left corner to the bottom right corner of the chessboard so that all of the rooks are on the same side of the path. In how many different ways can Claudia complete this task? (Two different paths for the same placement of rooks are considered as different.)

Solution: We call a placement of the rooks paired with a path satisfying the given conditions a *situation*. It is important to note that it is possible to have more than one path paired with a certain placement, giving different situations. By symmetry, we need to consider only the situations in which all of the rooks are under the path; that is, the bottom left corner and the rooks are on the same side of the path. Let S_n denote the set of such situations. Then the answer to the problem is $2|S_n|$. We claim that $|S_n| = (2n-1)|S_{n-1}|$, and the answer to the problem is $2\prod_{i=1}^{n}(2i-1) = 2 \cdot (2n-1)!!$.

Consider a situation s_n in S_n. We can describe s_n using a $2 \times n$ matrix. More precisely, we write

$$s_n = \begin{bmatrix} a_1 & a_2 & a_3 & \ldots & a_n \\ b_1 & b_2 & b_3 & \ldots & b_n \end{bmatrix},$$

where for each $1 \leq i \leq n$, a_i denotes the number of unit squares of the chessboard that are under the path in the ith column (from the left), and the rook in the ith column is placed in the b_i^{th} square (from the bottom). From the given conditions, we conclude that a_i and b_i are positive integers such that (1) $n \geq a_1 \geq a_2 \geq \cdots \geq a_n \geq 1$; (2) $a_i \geq b_i$ for $1 \leq i \leq n$; and (3) $(b_1, b_2, \ldots, b_n)$ is a permutation of $(1, 2, \ldots, n)$. Let A_n denote the set of all such matrices. It is clear that there is a bijection between S_n and A_n. (Also note that $a_1 = n$.) Since $|S_1| = |A_1| = 2$, it suffices to show that $|A_n| = (2n - 1)|A_{n-1}|$. We now define a map from A_n to A_{n-1}. For a matrix s_n in A_n, there is a k such that $b_k = 1$. We map s_n to

$$s_{n-1} = \begin{bmatrix} a_1 - 1 & a_2 - 1 & \ldots & a_{k-1} - 1 & a_{k+1} - 1 & \ldots & a_n - 1 \\ b_1 - 1 & b_2 - 1 & \ldots & b_{k-1} - 1 & b_{k+1} - 1 & \ldots & b_n - 1 \end{bmatrix}.$$

Because $b_k = 1$, we know that $b_i > 1$ for $i \neq k$, so that $n - 1 \geq a_i - 1 \geq b_i - 1 \geq 1$ and $(b_1 - 1, \ldots, b_{k-1} - 1, b_{k+1} - 1, \ldots, b_n - 1)$ is a permutation of $(1, 2, \ldots, n - 1)$. Thus s_{n-1} is in A_{n-1}. For example, the mapping shown in Figure 5.4

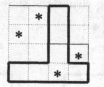

Figure 5.4.

corresponds to

$$s_4 = \begin{bmatrix} 4 & 4 & 2 & 2 \\ 3 & 4 & 1 & 2 \end{bmatrix} \rightarrow s_3 = \begin{bmatrix} 3 & 3 & 1 \\ 2 & 3 & 1 \end{bmatrix}.$$

It suffices to show that this mapping is $(2n-1)$-to-1; that is, for every s_{n-1} in A_{n-1} there are exactly $2n - 1$ matrices in A_n whose image is s_{n-1}. Let

$$s_{n-1} = \begin{bmatrix} c_1 & c_2 & c_3 & \ldots & c_{n-1} \\ d_1 & d_2 & d_3 & \ldots & d_{n-1} \end{bmatrix},$$

be an element in A_{n-1}. Then s_{n-1} is the image of

$$s_{n,k} = \begin{bmatrix} c_1+1 & c_2+1 & \cdots & c_k+1 & c & c_{k+1}+1 & \cdots & c_{n-1}+1 \\ d_1+1 & d_2+1 & \cdots & d_k+1 & 1 & d_{k+1}+1 & \cdots & d_{n-1}+1 \end{bmatrix},$$

where $c_k \mid 1 \geq c \geq c_{k+1}+1$ for $1 \leq k \leq n-2$, $c = n$ for $k = 0$ (that is, c is inserted in the leftmost column), and $c_{n-1} \geq c \geq 1$ for $k = n-1$ (that is, c is inserted in the rightmost column). In other words, for a fixed $0 \leq k \leq n-1$, there are $c_k - c_{k+1} + 1$ such values of c (where $c_0 = c_1 = n$ and $c_n = 1$). Hence, the total number of matrices in A_n that are mapped to s_{n-1} is equal to

$$\sum_{k=0}^{n-1}(c_k - c_{k+1} + 1) = \sum_{k=0}^{n-1}(c_k - c_{k+1}) + n = c_0 - c_n + n = 2n-1.$$

Consequently, we have $|A_n| = (2n-1)|A_{n-1}|$, as claimed. The following display illustrates the six elements besides s_4 that are mapped to s_3

$$\begin{bmatrix} 4 & 4 & 4 & 2 \\ 1 & 3 & 4 & 2 \end{bmatrix}, \quad \begin{bmatrix} 4 & 4 & 4 & 2 \\ 3 & 1 & 4 & 2 \end{bmatrix}, \quad \begin{bmatrix} 4 & 4 & 4 & 2 \\ 3 & 4 & 1 & 2 \end{bmatrix},$$

$$\begin{bmatrix} 4 & 4 & 3 & 2 \\ 3 & 4 & 1 & 2 \end{bmatrix}, \quad \begin{bmatrix} 4 & 4 & 2 & 1 \\ 3 & 4 & 2 & 1 \end{bmatrix}, \quad \begin{bmatrix} 4 & 4 & 2 & 2 \\ 3 & 4 & 2 & 1 \end{bmatrix}.$$

This corresponds to the following situations shown in Figure 5.5.

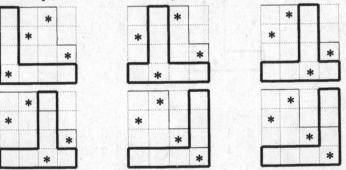

Figure 5.5.

In order to solve such a difficult problem, it is always helpful to calculate some initial values. For example, it is easy to see that $|S_1| = 1$ and $|S_2| = 3$ (Figure 5.6).

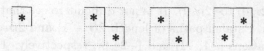

Figure 5.6.

Without too much trouble, we can also find $|S_3| = 15$, because there are $3! = 6$, $2 \cdot 2 = 4$, 2, 2, and 1 possible placements of rooks, respectively, for each of the following paths (Figure 5.7).

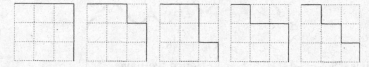

Figure 5.7.

From the initial values 1, 3, and 15, we might guess that $|S_n| = (2n - 1)!!$. But on the other hand, 1, 3, 15 are also the initial values of the sequence $\{2^{2^{n-1}} - 1\}_{n=1}^{\infty}$. While it takes some effort to obtain $|S_4| = 105$, this is certainly not impossible.   ■

The last problem is a combinatorial model of the identity

$$\sum (a_1 a_2 \cdots a_n) = \prod_{k=1}^{n} (2k - 1) = (2n - 1)!!,$$

where the sum is taken over all ordered n-tuples of integers $(a_1, \ldots, a_n)$ with $1 \leq a_i \leq n + 1 - i$ and $a_{i+1} \geq a_i - 1$ for all i. Can you see why this is true?

We will use the next two examples to revisit the Catalan numbers. Let $\mathbf{u}$ and $\mathbf{v}$ denote the vectors $[1, 1]$ and $[1, -1]$, respectively. Let n be a positive integer. A **Catalan path of length** $2n$ is a path in the coordinate plane using n $\mathbf{u}$'s and n $\mathbf{v}$'s to connect $(0, 0)$ and $(2n, 0)$ without crossing the x-axis.

Example 5.13. Determine the number of Catalan paths of length $2n$.

Solution: Let A_n denote the set of Catalan paths of length $2n$, and let $a_n = |A_n|$. We partition the paths in A_n into n subsets. For a path $\mathcal{P}_n$ in A_n, we put $\mathcal{P}_n$ into subset S_r if the point $(2r, 0)$ is the leftmost positive x-intercept of $\mathcal{P}_n$. If $\mathcal{P}_n$ is an element of S_r, then we can decompose $\mathcal{P}_n$ into four parts: (1) a $\mathbf{u}$ starting at $(0, 0)$; (2) a path $\mathcal{P}_{r-1}$ from $(1, 1)$ to $(2r - 1, 1)$; (3) a $\mathbf{v}$ ending at $(2r, 0)$; (4) a path

$\mathcal{P}_{n-r}$ from $(2r, 0)$ to $(2n, 0)$. It is not difficult to see that paths $\mathcal{P}_{r-1}$ and $\mathcal{P}_{n-r}$ are Catalan paths of length $2(r-1)$ and $2(n-r)$ via the vector translations $[-1, -1]$ and $[-2r, 0]$, respectively. For example, Figure 5.8 shows the decomposition of a Catalan path of length 16. In this path, $r = 5$, and the two darkened paths are Catalan paths of length 8 and 6 via the vector translations $[-1, -1]$ and $[-10, 0]$, respectively.

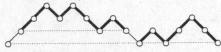

Figure 5.8.

Thus, $|S_r| = |A_{r-1}| \cdot |A_{n-r}|$. Hence we have

$$|A_n| = |S_1| + |S_2| + \cdots + |S_n|$$

$$= 1 \cdot |A_{n-1}| + |A_1| \cdot |A_{n-2}| + \cdots + |A_{n-2}| \cdot |A_1| + |A_{n-1}| \cdot 1,$$

that is, $a_n = |A_n|$ satisfies the recursion

$$a_n = a_0 a_{n-1} + a_1 a_{n-2} + \cdots + a_{n-1} a_0, \qquad (\ddagger)$$

where we set $a_0 = 1$. We will postpone solving this recursion to Chapter 8, where we discuss generating functions.

But why are these called Catalan paths? Readers might already see the relation between Examples 4.11 and 5.13. If we consider $\mathbf{u}$ as a \$5 bill and $\mathbf{v}$ as a \$10 bill, it is easy to see that these two problems are the same. Thus $|A_n| = \frac{1}{n+1}\binom{2n}{n}$, and the Catalan numbers satisfy the recursive relation $\ddagger$.   ∎

Example 5.14. Let n be a positive integer. Zachary is given $2n$ of each of vectors $\mathbf{u}$ and $\mathbf{v}$. He is going to draw a path using the $4n$ vectors to connect $(0, 0)$ and $(4n, 0)$ without passing through any of the points $(2, 0)$, $(6, 0)$, $\ldots$, or $(4n - 2, 0)$. In how many ways can Zachary accomplish this?

First Solution: Let Z_{2n} denote the set of all paths that Zachary can draw. We will partition Z_{2n} into n subsets. For a path $\mathcal{P}_{2n}$ in Z_{2n}, we put $\mathcal{P}_{2n}$ into subset T_r if the point $(4r, 0)$ is the leftmost positive x-intercept of $\mathcal{P}_{2n}$. If $\mathcal{P}_n$ is an element of S_r, then we can decompose $\mathcal{P}_{2n}$ into four parts: (1) a $\mathbf{u}$ starting at $(0, 0)$ (or a $\mathbf{v}$ starting at $(0, 0)$); (2) a path $\mathcal{P}_{2r-1}$ from $(1, 1)$ to $(4r - 1, 1)$ (or from $(1, -1)$ to $(4r - 1, -1)$); (3) a $\mathbf{v}$ ending at $(4r, 0)$ (or a $\mathbf{u}$ ending at

$(4r, 0)$; (4) a path $\mathcal{P}_{2n-2r}$ from $(4r, 0)$ to $(4n, 0)$. It is easy to see that $\mathcal{P}_{2r-1}$ is a Catalan path of length $4r - 2$ via the vector translation $[-1, -1]$ (or a vector translation $[-1, 1]$ followed by a reflection across the x-axis). It is also easy to see that $\mathcal{P}_{2n-2r}$ is in Z_{2n-2r} via the vector translation $[-4r, 0]$. Let $z_{2n} = |Z_{2n}|$ and $a_n = |A_n|$, where A_n denotes the set of all Catalan paths of lengths $2n$. We have

$$z_{2n} = 2[a_1 z_{2n-2} + a_3 z_{2n-4} + \cdots + a_{2n-1}]. \qquad (\ddagger')$$

This recursion looks somehow similar to the recursion $(\ddagger)$. Indeed, we claim that $z_{2n} = a_{2n}$. We induct on n. We set $z_0 = a_0 = 1$. We can rewrite recursion $(\ddagger')$ as

$$z_{2n} = 2(a_1 z_{2n-2} + a_3 z_{2n-4} + \cdots + a_{2n-1} z_0).$$

It is easy to count directly that $z_2 = 2 = a_2$. Then

$$z_4 = 2(a_1 z_2 + a_3 z_0) = 2(a_1 a_2 + a_3 a_0) = a_0 a_3 + a_1 a_2 + a_2 a_1 + a_3 a_0 = a_4$$

by recursion $(\ddagger)$. Assume that $z_{2n} = a_{2n}$ for $n = 0, 1, 2, \ldots, k-1$. Then

$$z_{2k} = 2[a_1 a_{2k-2} + a_3 a_{2k-4} + \cdots + a_{2k-1} a_0] = a_{2k},$$

by recursion $(\ddagger)$. Our induction is complete.   ∎

Instead of using the above approach, we could also set up a bijection between A_{2n} and Z_{2n}.

Second Solution: Our bijection will take each path in Z_{2n} to a path in A_{2n} that has the same x-intercepts. Each path in Z_{2n} is divided into consecutive legs that begin and end at points $(4k, 0)$ and stay completely above or below the x-axis in between. We leave fixed any leg that stays above the x-axis. As for legs that stay below, we map them to the remaining possible legs for paths in A_{2n}. Those legs with nonnegative y-coordinates that were not counted because they hit the x-axis in at least one point of the form $(4k - 2, 0)$ for some positive integer k.

Thus, for a fixed leg-length of $4m$, without loss of generality going from $(0, 0)$ to $(4m, 0)$, we need to display a bijection between the set of paths that go from $(0, 0)$ to $(4m, 0)$ staying strictly below the x-axis, and the set of paths from $(0, 0)$ to $(4m, 0)$ that stay above the x-axis, except at one or more points $(4k - 2, 0)$ where they intersect the x-axis. Notice that there is a bijection between the first set and

A_{2m-1}, because we can just remove the first and last vectors and then flip the path across the x-axis. Hence it suffices to solve the following problem:

> *Adrian is given $2m$ vectors **u** and $2m$ vectors **v**. He is going to draw a path by using the $4m$ vectors to connect $(0,0)$ and $(4m,0)$. Furthermore, the path does not intersect the line $y = -1$, while it intersects the x-axis at $(0,0)$, $(4m,0)$, $(4m_1 - 2, 0)$, $\ldots$, $(4m_k - 2, 0)$, where $m_1, \ldots, m_k$ are integers with $1 \leq m_1 < \cdots < m_k \leq m$. Show that Adrian can accomplish this in $|A_{2m-1}|$ ways.*

Let B_{2m} denote the set of all paths that Adrian can draw. It suffices to show that there is a bijection between B_{2m} and A_{2m-1}. Indeed, we set up the bijection as follows: For any path $\mathcal{P}$ in A_{2m-1}, consider its greatest x-intercept with x-coordinate divisible by 4; that is, the intercept can be written in the form $(4r, 0)$ for some nonnegative integer r. Note that at least one such point exists, because $(0,0)$ is such a point. Then all intercepts beyond $(4r, 0)$ are of the form $(4k - 2, 0)$. Let $\mathcal{P}_1$ denote the part of $\mathcal{P}$ from $(0,0)$ to $(4r, 0)$, and let $\mathcal{P}_2$ denote the part of $\mathcal{P}$ from $(4r, 0)$ to $(4m - 2, 0)$. For example, Figure 5.9 shows a Catalan path in A_9 with $m = 5$ and $r = 2$. The darkened part of the path indicates $\mathcal{P}_2$.

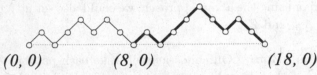

(0, 0)   *(8, 0)*   *(18, 0)*

Figure 5.9.

Now we take $\mathcal{P}_2$ and translate it $4r$ units to the left to obtain path $\mathcal{B}_1$ connecting $(0,0)$ and $(4m - 4r - 2, 0)$. Observe that the x-intercepts of $\mathcal{B}_1$ are all of the form $(4k - 2, 0)$, as desired, and that there exists at least one such intercept, namely, $(4m - 4r - 2, 0)$, the right end point of $\mathcal{B}_\infty$. We construct a path $\mathcal{B}_\in$ by first attaching **u** and **v** to $(4m - 4r - 2, 0)$ and $(4m, 0)$, respectively, then filling the gap from $(4m - 4r - 1, 1)$ to $(4m - 1, 1)$ by translating $\mathcal{P}_1$ by the vector $[4m - 4r - 1, 1]$. Because $\mathcal{P}_1$ does not intersect the line $y = -1$, the path $\mathcal{B}_2$ is strictly above the x-axis. Combining the paths $\mathcal{B}_1$ and $\mathcal{B}_2$ gives a path $\mathcal{B}$ in the set B_{2m} as the image of $\mathcal{P}$. (Figure 5.10 shows the image of the path in Figure 5.9, with the darkened part indicating $\mathcal{B}_1$.)

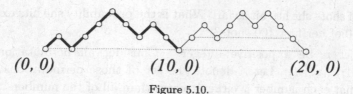

(0, 0)   (10, 0)   (20, 0)

Figure 5.10.

It is not difficult to see that this relation is indeed a bijection, because we have applied only vector translations. ∎

Exercises 5

5.1. There are n lines in a plane. Find the maximum number of
   (i) regions (open and closed) in the plane formed by the lines;
   (ii) closed regions in the plane bounded by the lines.

5.2. Zachary wants to write down all n-digit positive integers using the digits 1, 2, and 3, such that no two 1's are next to each other. How many such integers can Zachary write?

5.3. Let k and n be positive integers, and let $S = \{1, 2, \ldots, n\}$. A subset of S is called *skipping* if it does not contain consecutive integers. How many k-element skipping subsets are there? How many skipping subsets are there altogether? Find another approach to solve Example 3.8.

5.4. [Tower of Hanoi] We are given a small board into which three rods have been inserted and a set of n disks of different diameters with holes so that they can fit over the rods. Initially, all disks are on the same rod, with the largest disk on the bottom, the second largest disk above the largest disk, the third largest disk above the second largest disk, and so on. Adrian is asked to move the tower to one of the other rods in such a way that during the entire process no disk is above a smaller disk on the same rod. What is the minimum number of moves that Adrian has to make? (Adrian can move only one disk per move.)

5.5. [AIME 1990] A fair coin is to be tossed ten times. What is the probability that heads never occurs on consecutive tosses?

5.6. [Putnam 2002] Shanille O'Keal shoots free-throws on a basketball court. She hits the first and misses the second, and thereafter the probability that she hits the next shot is equal to the proportion of shots she has hit so far. What is the probability she hits exactly 50 of her first 100 shots?

5.7. Let n be a positive integer. Consider all permutations of $\{1, 2, \ldots, n\}$. Let A denote the set of those permutations such that each number is either greater than all of the numbers to its left or less than all of the numbers to its right. Let B denote the set of those permutations $a_1, a_2, \ldots, a_n$ such that for $1 \le i \le n-1$, there is a $j > i$ with $|a_j - a_i| = 1$. Show that $|A| = |B|$, and fill in the gaps in the solutions of Example 5.8.

5.8. Let $\{D_n\}_{n=1}^{\infty}$ be a sequence with $D_1 = 0$, $D_2 = 1$, and $D_{n+1} = n(D_n + D_{n-1})$ for all positive integers n. Express D_n in closed form.

5.9. [ARML 2003] ARMLovian, the language of the fair nation of ARMLovia, consists only of words using the letters A, R, M, and L. All words can be broken up into syllables that consist of exactly one vowel, possibly surrounded by a single consonant on either or both sides. For example, LAMAR, AA, RA, MAMMAL, MAMA, AMAL, LALA, MARLA, RALLAR, and AAALAAAAAMA are ARMLovlian words, but MRLMRLM, MAMMMAL, MMMMM, L, ARM, ALARM, LLAMA, and MALL are not. Compute the number of seven-letter ARMLovian words.

5.10. [China 1991] For any positive integer n, prove that the number of positive integers using only the digits 1, 3 and 4, whose digit-sum is $2n$ is a perfect square.

5.11. Each unit square of a $2 \times n$ unit square grid is to be colored blue or red such that there no 2×2 red square is obtained. Let c_n denote the number of different colorings. Determine, with proof, the greatest integer k such that 3^k divides c_{2001}.

5.12. [IMO 1997 Short–listed] Let n be a positive integer. In Exeter there are n girls and n boys, and each girl knows each boy. In Lincoln there are n girls $g_1, g_2, \ldots, g_n$, and $2n - 1$ boys $b_1, b_2, \ldots, b_{2n-1}$. Girl g_i, for $1 \le i \le n$, knows boys $b_1, b_2, \ldots, b_{2i-1}$ and no others. For all $r = 1, 2, \ldots, n$, denote by $A(r)$ and $B(r)$ the number of different ways in which r girls from Exeter and Lincoln, respectively, can dance with r boys from their own city to form r pairs, each girl with a boy she knows. Prove that $A(r) = B(r)$ for all r.

6
Inclusion-Exclusion Principle

The Inclusion–Exclusion Principle is helpful for counting the elements of the union of overlapping sets. We consider a few basic examples.

Example 6.1. Claudia wants to put five different candy bars into three different pockets so that no pocket is empty. In how many ways can this be done?

First Solution: Either two pockets each have two candy bars and the other pocket has one, or two pockets each have one candy bar and the other pocket has three. The first of these scenarios can occur in $3 \cdot 5 \cdot \binom{4}{2} = 90$ ways (there are 3 ways to choose the pocket with only one candy bar, 5 ways to choose that unique candy bar, and $\binom{4}{2}$ ways to choose, from the remaining four candy bars, the two that go in the first pocket with two candy bars). The second scenario can occur in $3 \cdot 5 \cdot 4 = 60$ ways (there are 3 ways to choose the pocket with three candy bars, 5 ways to choose the candy bar in the first pocket with one candy bar, and 4 ways to choose the candy bar in the other pocket with one candy bar). Thus, there are $90 + 60 = 150$ ways Claudia can arrange the candy bars.

Second Solution: Each of the five candy bars can go into any of the three pockets, so at first we might be tempted to think that the answer is $3^5 = 243$.

However, no pockets are allowed to be empty. If the first pocket is empty, then there are $2^5 = 32$ ways to arrange the candy bars in the remaining two pockets. Similarly, there are 32 arrangements in which the second pocket is empty, and 32 arrangements in which the third pocket is empty. Now we might be tempted to think that the answer is $243 - 32 - 32 - 32 = 147$.

But what if both the first pocket and the second pocket are empty? We subtracted this scenario once along with the first 32, and again with the second 32. But we really need to subtract it only once, so we need to add 1 back. Similarly, we need to add back another 1 to account for the case in which both the second and third pockets are empty, and yet another 1 for the case in which both the first and third pockets are empty.

Our final answer is therefore $147 + 1 + 1 + 1 = 150$.   ∎

Here, the first solution might seem easier than the second one. However, the first solution involved considering various arrangements, so that it becomes unusable for larger numbers of pockets and candy bars. The second solution, on the other hand, is more applicable in general, and its advantages are emphasized when the numbers involved in the problem are larger.

Example 6.2. Determine the number of primes less or equal to 111.

Solution: We want to find a systematic method rather than just list all primes that are less than 111.

Let t be the number of composite numbers in the set $A = \{1, 2, \ldots, 111\}$. Then we are looking for $111 - t - 1$ (since 1 is neither prime nor composite). If b is a factor of some $a \in A$, then $111 \geq a = bc$, where c is also an integer. Hence one of b or c is less than or equal to $\lfloor \sqrt{111} \rfloor$, where $\lfloor x \rfloor$ denotes the largest integer less than or equal to x. Thus all composite numbers in A are divisible by at least one of the primes $p_1 = 2, p_2 = 3, p_3 = 5$, and $p_4 = 7$. Let A_{p_i} be the subset of multiples of p_i in A. We have to determine the cardinality of the union of the four sets A_2, A_3, A_5, and A_7.

We count the composite numbers in the following manner. First we count all multiples of 2, then the multiples of 3, and so on. Since numbers of the form $p_i p_j m$ $(i \neq j)$, such as 6 and 10, are counted twice, we take out the multiples of 6, 10, 15, and so on. But then the numbers of the form $p_i p_j p_k m$, such as 30 and 42, are not counted at all, since each was counted 3 times in the first step and then was taken out three times in the second step, so we have to add the multiples of 30, 42, and so on. Finally, we have consider the numbers of the form $p_1 p_2 p_3 p_4 m$, which have been counted 4 times in the first step, taken out 6 times in the second step, and counted 4 times in the third step, so that we need to take them out one more time. Note that there are $\lfloor \frac{n}{m} \rfloor$ multiples of m in the set $\{1, 2, \ldots, n\}$. It follows that

$$
\begin{aligned}
&|A_2 \cup A_3 \cup A_5 \cup A_7| \\
&= \left\lfloor \frac{111}{2} \right\rfloor + \left\lfloor \frac{111}{3} \right\rfloor + \left\lfloor \frac{111}{5} \right\rfloor + \left\lfloor \frac{111}{7} \right\rfloor \\
&\quad - \left\lfloor \frac{111}{6} \right\rfloor - \left\lfloor \frac{111}{10} \right\rfloor - \left\lfloor \frac{111}{14} \right\rfloor - \left\lfloor \frac{111}{15} \right\rfloor - \left\lfloor \frac{111}{21} \right\rfloor - \left\lfloor \frac{111}{35} \right\rfloor \\
&\quad + \left\lfloor \frac{111}{30} \right\rfloor + \left\lfloor \frac{111}{42} \right\rfloor + \left\lfloor \frac{111}{70} \right\rfloor + \left\lfloor \frac{111}{105} \right\rfloor - \left\lfloor \frac{111}{210} \right\rfloor \\
&= 55 + 37 + 22 + 15 - 18 - 11 - 7 - 7 - 5 - 3 \\
&\quad + 3 + 2 + 1 + 1 - 0 \\
&= 85.
\end{aligned}
$$

Hence we find that the set $\{1, 2, \ldots, 111\}$ contains exactly $111 + 4 - 1 - 85 = 29$ prime numbers (here we add 4 because we have counted p_1, p_2, p_3, and p_4 as being composite). ∎

More generally, let n be a positive integer greater than 1. We want to determine the number of primes in the set $S = \{1, 2, \ldots, n\}$. Let us observe that every composite number in S is a multiple of at least one prime less than or equal to $\lfloor \sqrt{n} \rfloor$. Let $p_1 = 2, p_2 = 3, \ldots, p_k$ be the sequence of primes less than or equal to $\lfloor \sqrt{n} \rfloor$. For each p_i in this sequence we denote by A_{p_i} the subset of multiples of p_i in A. Then the number of primes less than or equal to n is

$$
n + k - 1 - \left| \bigcup_{i=1}^{k} A_{p_i} \right|.
$$

Generally speaking, we have the following technique, called the **Inclusion–Exclusion Principle** (also known as the **Boole–Sylvester Formula**).

Theorem 6.1. *Let $A_1, \ldots, A_n$ be a family of finite sets. Then the number of elements in the union $A_1 \cup \cdots \cup A_n$ is given by*

$$
\left| \bigcup_{i=1}^{n} A_i \right| = \sum_{\substack{I \subseteq \{1, \ldots, n\} \\ I \neq \emptyset}} (-1)^{|I|+1} \left| \bigcap_{i \in I} A_i \right|. \tag{$*$}
$$

Proof: We proceed by induction. For $n = 2$ we have to prove that

$$|A_1 \cup A_2| = |A_1| + |A_2| - |A_1 \cap A_2|.$$

This follows from the fact that $A_1 \cup A_2$ is the union of the disjoint sets A_1 and $A_2 \backslash (A_1 \cap A_2)$, while A_2 is the union of the disjoints sets $A_2 \backslash (A_1 \cap A_2)$ and $A_1 \cap A_2$. From the equalities

$$|A_2| = |A_2 \backslash (A_1 \cap A_2)| + |A_1 \cap A_2|,$$

$$|A_1 \cup A_2| = |A_1| + |A_2 \backslash (A_1 \cap A_2)|,$$

we obtain the desired equality.

Suppose the statement is true for families of $n - 1$ sets. We will prove it for families of n sets:

$$\left| \bigcup_{i=1}^{n} A_i \right| = \left| \bigcup_{i=1}^{n-1} A_i \cup A_n \right| = \left| \bigcup_{i=1}^{n-1} A_i \right| + |A_n| - \left| \left(\bigcup_{i=1}^{n-1} A_i \right) \cap A_n \right|$$

$$= \left| \bigcup_{i=1}^{n-1} A_i \right| + |A_n| - \left| \bigcup_{i=1}^{n-1} (A_i \cap A_n) \right|$$

$$= \sum_{\substack{I \subseteq \{1,\ldots,n-1\} \\ I \neq \emptyset}} (-1)^{|I|+1} \left| \bigcap_{i \in I} A_i \right| + |A_n|$$

$$- \sum_{\substack{I \subseteq \{1,\ldots,n-1\} \\ I \neq \emptyset}} (-1)^{|I|+1} \left| \bigcap_{i \in I} (A_i \cap A_n) \right|.$$

By grouping the sums having the same numbers of factors in their intersections, we obtain formula (∗).   ∎

A dual version of Theorem 6.1 is the following.

Theorem 6.2. *Let $A_1, \ldots, A_n$ be a family of finite subsets of the set S, and let $\overline{A_i} = S - A_i$ be the complement of A_i. Then*

$$\left| \bigcap_{i=1}^{n} \overline{A_i} \right| = |S| + \sum_{\substack{I \subseteq \{1,\ldots,n\} \\ I \neq \emptyset}} (-1)^{|I|} \left| \bigcap_{i \in I} A_i \right|.$$

Solution: Because $\bigcap_{i=1}^{n} \overline{A_i}$ and $\bigcup_{i=1}^{n} A_i$ form a partition of S, the desired result follows from Theorem 6.1.   ∎

A more general form of the Inclusion–Exclusion Principle is the following.

Theorem 6.3. *Let A be a finite set and f a function from A to the real numbers. For every subset $B \subseteq A$ we set*

$$f(B) = \sum_{x \in B} f(x),$$

where $f(\emptyset) = 0$. If $A = \bigcup_{i=1}^{n} A_i$, then

$$f(A) = \sum_{I \neq \emptyset} (-1)^{|I|+1} f\left(\bigcap_{i \in I} A_i\right). \qquad (**)$$

If f is the constant function, that is, $f(x) = 1$ for all $x \in A$, then $(**)$ becomes $(*)$. The proof of this more general result is similar to the proof of $(*)$. We leave it as an exercise.

We now present three *low-dimensional* examples, that is, examples that apply Theorem 6.1 when n is relatively small.

Example 6.3. [AIME 2001] Each unit square of a 3-by-3 unit-square grid is to be colored either blue or red. For each square, either color is equally likely. What is the probability of obtaining a grid that does not have a 2-by-2 red square?

Solution: Number the squares as shown in Figure 6.1.

1	2	3
4	5	6
7	8	9

Figure 6.1.

For $i=1$, 2, 4, and 5, let Q_i be the event that i is the upper left corner of a 2-by-2 red square, and let $p(E)$ be the probability that event E will occur. By the Inclusion–Exclusion Principle, the probability that the grid has at least one 2-by-2 red square is equal to

$$p(Q_1) + p(Q_2) + p(Q_4) + p(Q_5)$$
$$-p(Q_1 \cap Q_2) - p(Q_1 \cap Q_4) - p(Q_2 \cap Q_5)$$
$$-p(Q_4 \cap Q_5) - p(Q_1 \cap Q_5) - p(Q_2 \cap Q_4)$$
$$+p(Q_1 \cap Q_2 \cap Q_5) + p(Q_1 \cap Q_2 \cap Q_4)$$
$$+p(Q_1 \cap Q_4 \cap Q_5) + p(Q_2 \cap Q_4 \cap Q_5)$$
$$-p(Q_1 \cap Q_2 \cap Q_4 \cap Q_5),$$

or

$$4\left(\frac{1}{2}\right)^4 - \left[4\left(\frac{1}{2}\right)^6 + 2\left(\frac{1}{2}\right)^7\right] + 4\left(\frac{1}{2}\right)^8 - \left(\frac{1}{2}\right)^9 = \frac{95}{512}.$$

The probability that the grid does *not* have at least one 2-by-2 red square is therefore $1 - 95/512 = 417/512$.   ∎

Example 6.4.   [AIME 1996] A $150 \times 324 \times 375$ rectangular solid is made by gluing together $1 \times 1 \times 1$ cubes. How many $1 \times 1 \times 1$ cubes will an internal diagonal of this solid pass through?

Solution: Let the rectangular solid have width w, length l, and height h, where w, l, and h are positive integers. We will show that the diagonal passes through the interiors of

$$w + l + h - \gcd(w, l) - \gcd(l, h) - \gcd(h, w) + \gcd(w, l, h)$$

$1 \times 1 \times 1$ cubes, where $\gcd(x, y)$ denotes the greatest common divisor of x and y.

First orient the solid in 3-space so that one vertex is at $O = (0, 0, 0)$ and another is at $A = (w, l, h)$. Then $\overline{OA}$ is a diagonal of the solid. Let $P = (x, y, z)$ be a point on this diagonal. Exactly one of $x, y,$ or z is an integer if and only if P is interior to a face of one of the small cubes. Exactly two of $x, y,$ and z are integers if and only if P is on an edge of one of the small cubes. All three of $x, y,$ and z are integers if and only if P is a vertex of one of the small cubes. As P moves along the diagonal from O to A, it leaves the interior of a small cube precisely when at least one of the coordinates of P is a positive integer. Thus the number of interiors of small cubes through which the diagonal passes is equal to the number of points on the diagonal with at least one positive integer coordinate. Points with positive coordinates on $\overline{OA}$ have the form

$$P = (wt, lt, ht) \quad \text{with} \quad 0 < t \le 1.$$

The first coordinate, wt, will be a positive integer for w values of t, namely the values $t = 1/w, 2/w, 3/w, \ldots, w/w$. The second coordinate will be an integer for l values of t, and the third coordinate will be an integer for h values of t. The sum $w + l + h$ doubly counts the points with two integer coordinates, however, and it triply counts the points with three integer coordinates. The first two coordinates will be positive integers precisely when t has the form $k/\gcd(w, l)$ for some positive integer k between 1 and $\gcd(w, l)$, inclusive. A similar argument shows that the second and third coordinates will be positive integers for $\gcd(l, h)$ values of t, the third and first coordinates will be positive integers for $\gcd(h, w)$ values of t, and all three will be positive integers for $\gcd(w, l, h)$ values of t. By the Inclusion–Exclusion Principle, P will have one or more positive integer coordinates

$$w + l + h - \gcd(w, l) - \gcd(l, h) - \gcd(h, w) + \gcd(w, l, h)$$

times, which gives 768 when $\{w, l, h\} = \{150, 324, 375\}$.   ∎

Example 6.5.   Let n be a positive integer with $n \ge 2$. Fix $2n$ points in space in such a way that no four of them are in the same plane, and select any $n^2 + 1$ segments determined by the given points. Prove that these segments form at least n triangles.

Example 6.5 is a stronger version of Example 6.5a. The latter can be proved by induction on n, but we will leave this as an exercise. We use the claim of Example 6.5a in the proof of Example 6.5.

Example 6.5a.   Let n be a positive integer with $n \ge 2$. Fix $2n$ points in space in such a way that no four of them are in the same plane, and select any $n^2 + 1$ segments determined by the given points. Prove that these segments form at least one triangle.

Solution: We use induction on n. If $n = 2$, then 5 segments are drawn. There are at most $\binom{4}{2} = 6$ segments and they form $\binom{4}{3} = 4$ triangles. Note that each segment is part of two triangles. The removal of one of the 6 segments will break up exactly two of the 4 triangles. Hence there are 2 triangles formed by the 5 segments.

We assume that the statement in the problem is true for $n = 1, 2, \ldots, k$, where k is a positive integer. Now suppose we are given $2n = 2(k+1)$ points $P_1, P_2, \ldots, P_{2k+2}$, and that $(k+1)^2 + 1$ segments are constructed. By Example 6.5a, we know that three of these segments form a triangle. Without loss of generality, assume that the triangle is $P_1 P_2 P_3$. We say that points P_i and P_j ($1 \le i < j \le n$) are *connected* if $\overline{P_i P_j}$ is one of the constructed segments. Let $S = \{P_4, \ldots, P_{2k+2}\}$, and for $i = 1, 2, 3$, let S_i denote the elements in S that are connected to P_i. If P_k is in $S_i \cap S_j$, then $P_i P_j P_k$ is a triangle. Then the segments form at least $t_{k+1} = |S_1 \cap S_2| + |S_2 \cap S_3| + |S_3 \cap S_1|$ triangles. By the Inclusion–Exclusion Principle, we have

$$|S_1 \cup S_2 \cup S_3| = |S_1| + |S_2| + |S_3| - t_{k+1} + |S_1 \cap S_2 \cap S_3|,$$

implying that

$$t_{k+1} \ge |S_1| + |S_2| + |S_3| - (2k - 1),$$

because $|S_1 \cup S_2 \cup S_3| \le 2k + 2 - 3 = 2k - 1$ and $|S_1 \cap S_2 \cap S_3| \ge 0$.

If $|S_1| + |S_2| + |S_3| \ge 3k - 1$, then $t_{k+1} \ge k$; that is, there are at least k triangles besides triangle $P_1 P_2 P_3$, and we are done.

We assume that $|S_1| + |S_2| + |S_3| \le 3k - 1$, or

$$(|S_1| + |S_2|) + (|S_2| + |S_3|) + (|S_3| + |S_1|) \le 6k - 2.$$

Hence, by the **Pigeonhole Principle**, at least one of $|S_1| + |S_2|$, $|S_1| + |S_2|$, and $|S_1| + |S_2|$ is no greater than the average $\frac{6k-2}{3}$. Without loss of generality, we may assume that $|S_1| + |S_2| \le 2k - 1$ (because $|S_1| + |S_2|$ is an integer). If we remove P_1 and P_2, we lose at most $|S_1| + |S_2| + 1 = 2k$ segments. Therefore, there are at least $(k+1)^2 - 2k = k^2 + 1$ segments constructed among the $2k$ points $P_3, P_4, \ldots, P_{2k+2}$. By the induction hypothesis, the segments form at least k triangles. Thus we have at least $k + 1$ triangles in the given configuration of points $P_1, P_2, \ldots, P_{2k+2}$, and our induction is complete.   ∎

We now introduce some classical examples of the Inclusion–Exclusion Principle and some contest problems. As shown in Example 6.2, many basic Number Theory results are obtained by applying the Inclusion–Exclusion Principle. Let n be a positive integer. The **Euler (totient) function** $\phi(n)$ is defined to be the number of positive integers no greater than n that are relatively prime to n. By convention, we set $\phi(1) = 1$.

If $p > 1$ is a prime number, then $1, 2, \ldots, p-1$ are relatively prime to p; thus $\phi(p) = p - 1$. If $p > 1$ is a prime number and $r \geq 1$, then the value of $\phi(p^r)$ is equal to the number of terms of the sequence

$$1, 2, 3, \ldots, p, p+1, \ldots, 2p, \ldots, p^r$$

that are not divisible by p. The numbers divisible by p in this sequence are $p, 2p, 3p, \ldots, p^r$; hence

$$\phi(p^r) = p^r - p^{r-1} = p^r \left(1 - \frac{1}{p}\right).$$

Example 6.6. Let n be a positive integer. Determine a general formula for $\phi(n)$.

Solution: Let $n = p_1^{r_1} \ldots p_m^{r_m}$ be the prime factorization of n; that is, $p_1, p_2, \ldots, p_m$ are distinct primes, and $r_1, r_2, \ldots, r_m$ are positive integers. A number a from the set $A = \{1, 2, \ldots, n\}$ is relatively prime to n if and only if a is not divisible by any of the numbers $p_1, \ldots, p_m$. For $i = 1, 2, \ldots, m$, we denote by A_i the subset of A that contains all of the numbers that are divisible by p_i. It follows that

$$A \setminus \bigcup_{i=1}^{m} A_i$$

is the set of all elements in A that are relatively prime to n. Hence,

$$\phi(n) = n - \left| \bigcup_{i=1}^{m} A_i \right|.$$

By Theorem 6.1, we obtain

$$\phi(n) = n - \sum_{\substack{I \subseteq \{1, \ldots, m\} \\ I \neq \emptyset}} (-1)^{|I|+1} \left| \bigcap_{i \in I} A_i \right|.$$

Note that $\bigcap_{i \in I} A_i$ contains those elements in A that are divisible by the product $\prod_{i\in I} p_i$. Hence $\left|\bigcap_{i\in I} A_i\right| = n \prod_{i\in I} \frac{1}{p_i}$. We have

$$\phi(n) = n + \sum_{\substack{I\subseteq\{1,\dots,m\}\\ I\neq\emptyset}} (-1)^{|I|} \cdot n \prod_{i\in I} \frac{1}{p_i}$$

$$= n \left[1 + \sum_{\substack{I\subseteq\{1,\dots,m\}\\ I\neq\emptyset}} \prod_{i\in I} \left(\frac{-1}{p_i}\right) \right] = n \prod_{i=1}^{m} \left(1 - \frac{1}{p_i}\right).$$

∎

We have determined the number of integers in the set $S = \{1, 2, \dots, n\}$ that are relatively prime to n. What is the sum of these numbers? It is not difficult to see that the sum is equal to $\frac{n\phi(n)}{2}$, because the numbers can be paired up so that the sum of each pair is equal to n.

An interesting application of the Inclusion–Exclusion Principle is the following result of G. Szegő and G. Pólya from 1925.

Example 6.7. [Szegő & Pólya] Let $n > 1$ be an integer, and let $a_1, a_2, \dots, a_{\phi(n)}$ be the numbers in the set $A = \{1, 2, \dots, n\}$ that are relatively prime to n. Then

$$a_1^2 + a_2^2 + \dots + a_{\phi(n)}^2 = \frac{\phi(n)}{6} \left(2n^2 + (-1)^k p_1 \cdot \dots \cdot p_k\right),$$

where $p_1, p_2, \dots, p_k$ are the distinct prime divisors of n.

Solution: Let $f(x) = x^2$, $f(A) = \sum_{x\in A} f(x) = \sum_{x\in A} x^2$, and $f(\emptyset) = 0$. For $1 \leq i \leq k$, let A_i denote the subset of all elements in A that are divisible by p_i. Then

$$f(A) = \sum_{i=1}^{n} i^2 = \frac{n(n+1)(2n+1)}{6} = \frac{n^3}{3} + \frac{n^2}{2} + \frac{n}{6}.$$

We have

$$f(A_i) = p_i^2 + (2p_i)^2 + \dots + \left(\frac{n}{p_i} p_i\right)^2$$

$$= p_i^2 \left[\frac{1}{3}\left(\frac{n}{p_i}\right)^3 + \frac{1}{2}\left(\frac{n}{p_i}\right)^2 + \frac{1}{6}\left(\frac{n}{p_i}\right)\right]$$

for all $1 \leq i \leq k$, and

$$f(A_i \cap A_j) = (p_i p_j)^2 \left[\frac{1}{3} \left(\frac{n}{p_i p_j} \right)^3 + \frac{1}{2} \left(\frac{n}{p_i p_j} \right)^2 + \frac{1}{6} \left(\frac{n}{p_i p_j} \right) \right]$$

for all $1 \leq i < j \leq k$, and so on. By Theorem 6.3, we have

$$a_1^2 + a_2^2 + \cdots + a_{\phi(n)}^2 = f(A) - f \left(\bigcup_{i=1}^{k} A_i \right)$$

$$= \frac{n^3}{3} + \frac{n^2}{2} + \frac{n}{6} - \sum_{I \neq \emptyset} (-1)^{|I|+1} f \left(\bigcap_{i \in I} A_i \right)$$

$$= \frac{n^3}{3} + \frac{n^2}{2} + \frac{n}{6} + \sum_{I \neq \emptyset} (-1)^{|I|} f \left(\bigcap_{i \in I} A_i \right).$$

Note that

$$\sum_{I \neq \emptyset} (-1)^{|I|} f \left(\bigcap_{i \in I} A_i \right) = \sum_{\substack{I \subseteq \{1,\ldots,k\} \\ I \neq \emptyset}} (-1)^{|I|} \prod_{i \in I} p_i^2 \left[\frac{1}{3} \left(n \prod_{i \in I} \frac{1}{p_i} \right)^3 + \right.$$

$$\left. \frac{1}{2} \left(n \prod_{i \in I} \frac{1}{p_i} \right)^2 + \frac{1}{6} \cdot n \prod_{i \in I} \frac{1}{p_i} \right].$$

It follows that

$$a_1^2 + a_2^2 + \cdots + a_{\phi(n)}^2$$

$$= \frac{n^3}{3} \left(1 + \sum_{\substack{I \subseteq \{1,\ldots,k\} \\ I \neq \emptyset}} \prod_{i \in I} \left(\frac{-1}{p_i} \right) \right) + \frac{n^2}{2} \left(1 + \sum_{\substack{I \subseteq \{1,\ldots,k\} \\ I \neq \emptyset}} (-1)^{|I|} \right) +$$

$$\frac{n}{6} \left(1 + \sum_{\substack{I \subseteq \{1,\ldots,k\} \\ I \neq \emptyset}} \prod_{i \in I} (-p_i) \right)$$

$$= \frac{n^3}{3} \prod_{i=1}^{k} \left(1 - \frac{1}{p_i} \right) + \frac{n^2}{2} \prod_{i=1}^{k} (1 - 1) + \frac{n}{6} \prod_{i=1}^{k} (1 - p_i).$$

By Example 6.6, we have

$$a_1^2 + a_2^2 + \cdots + a_{\phi(n)}^2 = \frac{n^2\phi(n)}{3} + \frac{\phi(n)\prod_{i=1}^{k}(-p_i)}{6}$$

$$= \frac{\phi(n)}{6}\left(2n^2 + (-1)^k p_1 \cdot \ldots \cdot p_k\right),$$

as desired.   ■

Example 6.8.   There are 10 seniors asking Mr. Fat to write college recommendation letters. Mr. Fat writes a letter for each one of them and then labels the 10 envelopes. If he randomly places one letter into each labeled envelope, what is the probability that no letter is placed in the correct envelope?

This is a modern version of the **Bernoulli–Euler Formula for misaddressed letters**. We consider the general case of n letters and n labelled envelopes. We label the envelopes $e_1, e_2, \ldots, e_n$ and the letters $1, 2, \ldots, n$ in such a way that the letter i corresponds to the envelope e_i. There are $n!$ ways to put the letters into the envelopes, because we can arrange the letters in any order and then place them into the envelopes $e_1, e_2, \ldots, e_n$ in that order. All such orders of the letters are permutations of $S = \{1, 2, \ldots, n\}$. We are interested in permutations π of S such that $\pi(i) \neq i$ for all $1 \leq i \leq n$, because letter $\pi(i)$, instead of letter i, will be placed in envelope e_i. If π is a permutation of S, we say that π has a **fixed point** if $\pi(i) = i$ for some $1 \leq i \leq n$. A permutation without a fixed point is called a **derangement**. It suffices to count the number of derangements of S. Let D_n denote this number.

Solution: For every i, $1 \leq i \leq n$, we denote by F_i the set of permutations for which i is a fixed point. Thus

$$D_n = n! - \left| \bigcup_{i=1}^{n} F_i \right|.$$

Note that $|F_i| = (n-1)!$, $|F_i \cap F_j| = (n-2)!$ (because there are $(n-2)!$ permutations left for $S - \{i, j\}$ after we fix i and j), and so on. By Theorem 6.1, we obtain

$$D_n = n! - \sum_{I \neq \emptyset} (-1)^{|I|+1} \left| \bigcap_{i \in i} F_i \right| = n! - \sum_{k=1}^{n} \binom{n}{k} (-1)^{k+1} (n-k)!$$

$$= \sum_{k=0}^{n} (-1)^k \binom{n}{k} (n-k)! = n! \sum_{k=0}^{n} \frac{(-1)^k}{k!}.$$

Setting $n = 10$ yields the number of ways in which Mr. Fat can misplace all of the letters. The probability of this happening is $\sum_{k=0}^{10} \frac{(-1)^k}{k!}$.   ∎

We note that

$$\lim_{n \to \infty} \frac{D_n}{n!} = \lim_{n \to \infty} \left(1 - \frac{1}{1!} + \frac{1}{2!} - \cdots + (-1)^n \frac{1}{n!} \right) = \frac{1}{e}.$$

It is also interesting to notice the following recursive formula, which can be derived from the above formula by straightforward computation:

$$D_{n+1} = (n+1)D_n + (-1)^{n+1} = nD_n + (D_n - (-1)^n) = n(D_n + D_{n-1}).$$

We can also obtain the above recursion directly by using bijections. In a derangement π of S, we have two choices:

(i) $\pi(1) = k$ and $\pi(k) = 1$ for some $1 \leq k \leq n$. Then there are $(n-1)$ possible values for k, and for each k there are D_{n-2} derangements for the remaining $n-2$ elements. Hence there are $(n-1)D_{n-2}$ such derangements.

(ii) $\pi(1) = k$ and $\pi(k) = m$ for some $1 < k, m \leq n$. Note that $k \neq m$. Then the elements $\pi(2), \pi(3), \ldots, \pi(n)$ are simply a derangement π' of $(2, \cdots, k-1, 1, k+1, \cdots, n)$, while $\pi(1) = k$. Again, there are $n-1$ possible values for k, and for each k there are D_{n-1} derangements. Hence there are $(n-1)D_{n-1}$ such derangements.

Therefore, $D_n = (n-1)(D_{n-2} + D_{n-1})$. Note that $D_1 = 0$ and $D_2 = 1$. We can obtain the explicit formula $D_n = n! \sum_{k=0}^{n} \frac{(-1)^k}{k!}$ from this recursion (as in Exercise 5.8).

Two closely related contest problems are the following.

Example 6.9. [IMO 1989] A permutation $(x_1, x_2, \ldots, x_{2n})$ of the set $\{1, 2, \ldots, 2n\}$, where n is a positive integer, is said to have property P if

$$|x_i - x_{i+1}| = n$$

for at least one $i \in \{1, 2, \ldots, 2n - 1\}$. Show that, for each n, there are more permutations with property P than without.

We again present two arguments. The first solution applies the Inclusion–Exclusion Principle, and the second uses a bijection, a general counting principle that we emphasized in chapter 4.

First Solution: In view of property P, consider the n pairs of integers

$$\{1, n+1\}, \ \{2, n+2\}, \ \cdots, \ \{n, 2n\}.$$

The number of permutations of $\{1, 2, \ldots, 2n\}$ in which any given k of the pairs are found as adjacent entries is $(2n - k)! 2^k$. (The k chosen pairs each have 2 internal orderings, and with the $2n - 2k$ remaining integers there are in total $2n - k$ objects to permute arbitrarily.) Therefore, by the Inclusion–Exclusion Principle, the number of permutations in which none of the n pairs occur as adjacent entries is

$$N = (2n)! - \binom{n}{1}(2n-1)! 2^1 + \binom{n}{2}(2n-2)! 2^2 - \cdots$$

$$\leq (2n)! - \binom{n}{1}(2n-1)! 2^1 + \binom{n}{2}(2n-2)! 2^2$$

$$= \binom{n}{2}(2n-2)! 2^2.$$

The inequality above follows from **Bonferonni's Inequalities**: If the truncated Inclusion–Exclusion formula stops after a plus sign, we obtain an expression that is too large, but if it stops after a minus sign, we obtain an expression that is too small.

It follows that

$$\frac{N}{(2n)!} \leq \frac{\binom{n}{2}(2n-2)! 2^2}{(2n)!} = \frac{n-1}{2n-1} < \frac{1}{2};$$

that is, fewer than half of the permutations of $\{1, 2, \ldots, 2n\}$ do not have property P.

Second Solution: For brevity, we will say that a permutation $X = (x_1, x_2, \ldots, x_{2n})$ has property P_i when $|x_i - x_{i+1}| = n$.

For $n = 1$, there are no permutations without the property P, so let assume $n \geq 2$. We prove the desired result by establishing a one-to-one correspondence between permutations without the property P and a proper subset of those with property P. Specifically, if X is a permutation without property P, then $x_{m+1} = x_1 \pm n$ for some unique $m \geq 2$, and we may associate X with the permutation

$$Y = f(X) = (x_2, \ldots, x_m, x_1, x_{m+1}, \ldots, x_{2n}).$$

Since X does not have property P_i, permutation Y has property P_m but no other property P_i.

Conversely, it is not difficult to see that if Y is any permutation that has property P_m for exactly one m, with $m \geq 2$, then there is a unique permutation X for which $f(X) = Y$. Thus, f is a one-to-one mapping of all permutations without property P onto the set of permutations that have property P_m for exactly one $m \geq 2$. This is certainly a proper subset of the permutations with property P, and the proof is complete.   ∎

Example 6.10. [China 2000, by Chun Su] The sequence $\{a_n\}_{n \geq 1}$ satisfies the conditions $a_1 = 0, a_2 = 1$, and

$$a_n = \frac{1}{2}na_{n-1} + \frac{1}{2}n(n-1)a_{n-2} + (-1)^n\left(1 - \frac{n}{2}\right),$$

for $n \geq 3$. Express

$$f_n = a_n + 2\binom{n}{1}a_{n-1} + 3\binom{n}{2}a_{n-2} + \cdots + n\binom{n}{n-1}a_1$$

in closed form.

Solution: It is straightforward to show by induction that

$$a_n = na_{n-1} + (-1)^n,$$

which implies

$$a_n = n! - \frac{n!}{1!} + \frac{n!}{2!} - \frac{n!}{3!} + \cdots + (-1)^n\frac{n!}{n!}.$$

Therefore, by Example 6.8, $a_n = D_n$.

Then f_n can be interpreted as follows: For each non-identity permutation of $(1, 2, \ldots, n)$, add one point then for each of these permutations, add one point for each fixed point of the permutation. Then f_n is the total number of points, scored by all of the non-identity permutations.

On the other hand, the total number of points can also be calculated as the sum of the points of each element gained in the non-identity permutations. There are $n! - 1$ non-identity permutations, and each number is fixed in $(n-1)! - 1$ non-identity permutations, for a total of

$$f_n = n! - 1 + n((n-1)! - 1) = 2 \cdot n! - n - 1$$

points.   ∎

In this solution, we counted a certain type of object, the total points, in two ways to set up an identity. We will discuss this technique in the next chapter. Now let us look at another kind of derangement.

Example 6.11. At the beginning of the 2002 season, the 20 players of Mr. Fat's soccer team lined up in a row to take a team picture. At the end of their undefeated season, the players decided to take another team picture so that everyone had a different right-hand neighbor in the new picture. In how many ways could they have done this?

Solution: Let Q_n denote the number of such lineups for n players. We label the players $1, 2, \ldots, n$ from left to right in the original picture (so that person i, $1 \le i \le n-1$, is the right-hand neighbor of person $i+1$ in the original lineup). Then Q_n is the number of permutations $\pi = \{\pi(1), \pi(2), \ldots, \pi(n)\}$ of $S = \{1, 2, \ldots, n\}$ such that $\pi(i+1) - \pi(i) \neq 1$ for $1 \le i \le n-1$.

Let A_i denote the set of permutations of S such that $\pi(i+1) - \pi(i) = 1$. An element of A_i is in bijective correspondence with the permutations of the $(n-1)$-element set $\{1, 2, \ldots, (i, i+1), \ldots n\}$, so there are $(n-1)!$ such permutations; that is, $|A_i| = (n-1)!$. For $1 \le i < j - 1 \le n - 2$, an element in $A_i \cap A_j$ is in bijective correspondence with the permutations of the $(n-2)$-element set $\{1, 2, \ldots, (i, i+1), \ldots, (j, j+1), \ldots n\}$, so there are $(n-2)!$ such permutations. If $1 \le i = j - 1 \le n - 2$, then an element in $A_i \cap A_j = A_i \cap A_{i+1}$ is in a bijective correspondence with the permutations of the $(n-2)$-element set $\{1, 2, \ldots, (i, i+1, i+2), \ldots n\}$, so there are $(n-2)!$ such permutations. In any case, we have $|A_i \cap A_j| = (n-2)!$. In exactly the same way, we can show that

$$|A_{i_1} \cap A_{i_2} \cap \cdots \cap A_{i_k}| = (n-k)!,$$

for $1 \leq i_1 < i_2 < \cdots < i_k \leq n-1$. Hence, by Theorem 6.1,

$$Q_n = n! - \sum_{I \neq \emptyset} (-1)^{|I|+1} \left| \bigcap_{i \in I} A_i \right|$$

$$= n! - \sum_{k=1}^{n-1} \binom{n-1}{k} (-1)^{k+1} (n-k)!$$

$$= (n-1)! \sum_{k=0}^{n-1} \frac{(-1)^k (n-k)}{k!},$$

or

$$Q_n = (n-1)! \left(n - \frac{n-1}{1!} + \frac{n-2}{2!} - \cdots + (-1)^{n-1} \frac{1}{(n-1)!} \right).$$

Setting $n = 20$ yields the answer to the original question.   ∎

By straightforward calculation, we have the identity

$$Q_n = D_n + D_{n-1}.$$

It may be of interest to find a bijection to prove the above identity.

We now present three famous results to close this chapter.

The following identity is due to Euler. Let m and n be positive integers with $m \leq n$. Then

$$\sum_{k=0}^{m} (-1)^k \binom{m}{k} (m-k)^n = \begin{cases} 0 & \text{if } n < m, \\ n! & \text{if } n = m. \end{cases}$$

The left-hand side of the identity mimics the format of the Inclusion–Exclusion Principle. Indeed, the identity is a special case of the following model.

Example 6.12. Let A and B be two finite sets, with $|A| = n$ and $|B| = m$. Determine the number $s_{n,m}$ of surjective functions $f : A \to B$.

Solution: Let $B = \{1, 2, \ldots, m\}$. Let B^A denote the set of all functions $f : A \to B$. For every $i, 1 \leq i \leq m$, denote by B_i the set of function $f : A \to B$ such that $i \notin f(A)$. The set of surjective functions $f : A \to B$ is

$$B^A \backslash \bigcup_{i=1}^{m} B_i.$$

Notice that $|B^A| = m^n, |B_i| = (m-1)^n, |B_i \cap B_j| = (m-2)^n$, and so on. Applying the Inclusion–Exclusion Principle, we obtain

$$s_{n,m} = m^n - \sum_{I \neq \emptyset} (-1)^{|I|+1} \left| \bigcap_{i \in I} B_i \right|$$

$$= m^n - \sum_{I \neq \emptyset} (-1)^{|I|+1} (m - |I|)^n$$

$$= m^n - \sum_{k=1}^{m} \binom{m}{k} (-1)^{k+1} (m-k)^n$$

$$= \sum_{k=0}^{m} (-1)^k \binom{m}{k} (m-k)^n.$$

If $m = n$, the surjective functions $f : A \to B$ are exactly the bijective ones, so the number of those functions is $n!$. If $m < n$, there are no surjective functions $f : A \to B$. Therefore, we have Euler's identity

$$\sum_{k=0}^{m} (-1)^k \binom{m}{k} (m-k)^n = s_{n,m} = \begin{cases} 0 & \text{if } n < m, \\ n! & \text{if } n = m. \end{cases}$$

■

The following **Problème des ménages** is due to E. Lucas (1891).

Example 6.13. [Problème des ménages] How many ways can n married couples sit at a round table in such a way that there is one man between every two women and no man is seated next to his wife? (Rotations of the same configuration are considered to be equivalent.)

Solution: If we interpret this problem as a circular permutation problem, then the seats are treated as being indistinguishable. Let C_n be the answer to the problem under this assumption. There are $(n-1)!$ ways to seat all of the women. For each fixed arrangement of women, let M_n denote the number of ways to seat the men, with exactly one man between every two women and no man next to his wife. Then the answer to the problem is $C_n = (n-1)! M_n$.

On the other hand, we may assume that the seats are distinguishable. Let P_n denote the answer to the problem under this assumption. Then $P_n = 2n \cdot C_n$. Indeed, it is slightly easier to find P_n than to find M_n. Thus, we first assume that the seats are distinguishable. We label the seats $1, 2, \ldots, 2n$ in clockwise order. We label the women $w_1, w_2, \ldots, w_n$ and the men $m_1, m_2, \ldots, m_n$, such that m_i and w_i are married for $1 \le i \le n$.

Let A denote the set of seating arrangements in which each man is between two women. Then the men must take the seats in $\{1, 3, \ldots, 2n-1\}$ or in $\{2, 4, \ldots, 2n\}$. Hence $|A| = 2(n!)^2$. Let A_i denote the set of seating arrangements in A in which w_i and m_i are next to each other. Let $\overline{A_i} = A \backslash A_i$ be the complement of A_i. Then by Theorem 6.2,

$$P_n = \left| \bigcap_{i=1}^{n} \overline{A_i} \right| = |A| + \sum_{I \ne \emptyset} (-1)^{|I|} \left| \bigcap_{i \in I} A_i \right|.$$

By symmetry, it is easy to see that that if $|I| = |I'|$ for some $I, I' \subset \{1, 2, \ldots, n\}$, then $\left| \bigcap_{i \in I} A_i \right| = \left| \bigcap_{i \in I'} A_i \right|$. Therefore,

$$P_n = \left| \bigcap_{i=1}^{n} \overline{A_i} \right| = 2(n!)^2 + \sum_{k=1}^{n} (-1)^k \binom{n}{k} |B_k|,$$

where B_k denotes the seating arrangements in A in which w_i and m_i, $1 \le i \le k$, are next to each other. Now we count $|B_k|$. Let b_k be an element of B_k. We can obtain b_k in a few steps. First, all of the women have two choices: They can choose to sit in even-numbered seats or odd-numbered seats. Second, all of the couples w_i and m_i, $1 \le i \le k$, need to find k pairs of adjacent seats to sit in. Finally, when the pairs of seats are picked, there are $k!$ ways to seat the couples and $(n-k)!^2$ ways to seat everyone else. Thus

$$|B_k| = 2(k!)((n-k)!)^2 G_k,$$

where G_k denotes the number of ways to choose k pairs of seats. Consider the $2n$ pairs of seats $(1, 2), (2, 3), \ldots, (2n-1, n), (2n, 1)$ around the circle in clockwise order. Then G_k is equal to the number of ways to pick k pairs such that no two picked pairs are neighbors. This reduces the problem to Exercise 2.10 with $m = 2n$ and points $P_1, P_2, \ldots, P_m$ the pairs $(1, 2), (2, 3), \ldots, (n, 1)$, respectively:

Let m be a positive integer. Given m red points $P_1, P_2, \ldots, P_m$ placed around a circle in clockwise order, Adrian wants to chose k of the points and color them blue such that there are no neighboring blue points in the resulting configuration. In how many ways can Adrian accomplish his task?

Let g_m denote the number of coloring possibilities that Adrian has. We want to find $G_k = g_{2m}$. We consider $m - k$ boxes around the circle. We can classify Adrian's coloring possibilities into two categories depending on the color of P_m.

If point P_m is blue, then P_1 is red. In between the space from P_1 to P_m in the clockwise direction, we place $m - k - 1$ boxes. Among these boxes, $k - 1$ of them are of type X and $m - 2k$ are of type Y. A blue point and a red point are placed in clockwise order in a type X box. A red point is placed in a type Y box. (Thus P_m and P_1 would be in a box of type X.) There are $\binom{m-k-1}{k-1}$ ways to choose $k - 1$ boxes to be of type X, and each way of choosing those $k - 1$ boxes gives a unique coloring of the points.

If P_m is red, then we place $m - k$ boxes in the clockwise direction from P_m around the circle (the first box with P_m as its first entry). Among those boxes, k of them are of type X and $m - 2k$ are of type Y. There are $\binom{m-k}{k}$ ways to choose k boxes to be of type X, and each way of choosing those k boxes gives a unique coloring of the points.

Combining the two cases above gives

$$g_m = \binom{m-k-1}{k-1} + \binom{m-k}{k} = \left(\frac{k}{m-k} + 1\right) \cdot \binom{m-k}{k}$$

$$= \frac{m}{m-k}\binom{m-k}{k},$$

by Theorem 3.2 (d). Setting $m = 2n$ yields

$$G_k = \frac{2n}{2n-k}\binom{2n-k}{k}.$$

It follows that

$$P_n = 2(n!)^2 + \sum_{k=1}^{n}(-1)^k \binom{n}{k}|B_k|$$

$$= 2(n!)^2 + \sum_{k=1}^{n}(-1)^k \frac{n!}{(n-k)! \cdot k!} \cdot [2(k!)((n-k)!)^2 G_k]$$

$$= 2(n!)^2 + 2(n!)\sum_{k=1}^{n}(-1)^k \cdot ((n-k)!)G_k$$

$$= 2 \cdot n! \left[\sum_{k=0}^{n}(-1)^k \frac{2n}{2n-k} \cdot \binom{2n-k}{k}(n-k)!\right].$$

Therefore, the answer to the problem is

$$C_n = (n-1)! \left[\sum_{k=0}^{n} (-1)^k \frac{2n}{2n-k} \cdot \binom{2n-k}{k}(n-k)! \right].$$

∎

As a side result, we conclude that

$$M_n = \sum_{k=0}^{n} (-1)^k \frac{2n}{2n-k} \cdot \binom{2n-k}{k}(n-k)!.$$

Looking back at our solution, we realize that it is natural to use the Inclusion–Exclusion Principle to solve this problem. On the other hand, finding G_k is rather tricky. Yet we can approach the problem in a different way. Assume that Adrian has $m-k$ red points placed around a circle. He then puts k blue points in between the red points so that no two blue points are neighbors. Finally, he can label the points $P_1, P_2, \ldots, P_m$ in the resulting configuration. With some careful analysis of circular permutations, we can also obtain the formula for G_k.

Finally, we will present a result in counting matrices by M. Becheanu (1997).

Example 6.14. [Becheanu] Let p, m, and n be positive integers. Determine the number of m by n matrices with entries from the set $\{1, 2, \ldots, p\}$ which have the property that the sum of the elements in each row and each column is not divisible by p.

Solution: We denote by M the set of m by n matrices with entries from the set $\{1, 2, \ldots, p\}$. Let A_i be the subset of M formed by the matrices in which the sum of elements in the ith row is divisible by p, and B_j the subset of M consisting of matrices in which the sum of elements in the jth column is divisible by p. The desired number is

$$N = |M| - \left| \left(\bigcup_{i=1}^{m} A_i \right) \cup \left(\bigcup_{j=1}^{n} B_j \right) \right|$$

$$= \sum_{\substack{I \subset \{1,\ldots,m\} \\ J \subset \{1,\ldots,n\}}} (-1)^{|I|+|J|} \left| M \cap \bigcap_{i \in I} A_i \cap \bigcap_{j \in J} B_j \right|,$$

where $X \cap \bigcap_{i \in \emptyset} Y_i = X$. By symmetry, we have

$$N = \sum_{\substack{0 \leq k \leq m \\ 0 \leq \ell \leq n}} \binom{m}{k}\binom{n}{\ell}(-1)^{k+l} \left| M \cap \bigcap_{i=1}^{k} A_i \cap \bigcap_{j=1}^{\ell} B_j \right|.$$

In the case $0 \leq k < m$ and $0 \leq \ell \leq n$, if a matrix $A = (a_{i,j})$ belongs to $M \cap \bigcap_{i=1}^{k} A_i \cap \bigcap_{i=1}^{\ell} B_j$, then we can first pick the entries $a_{i,j}$, for $1 \leq i \leq k$ and $1 \leq j \leq n - 1$, arbitrarily from the set $\{1, 2, \ldots, p\}$. There are $p^{k \cdot (n-1)}$ ways to do so. Then the entries $a_{1,n}, a_{2,n}, \ldots, a_{k,n}$ are uniquely determined by the congruence relations

$$\sum_{j=1}^{n} a_{i,j} \equiv 0 \pmod{p} \quad \text{for all } 1 \leq i \leq k. \tag{$\dagger$}$$

We can then choose the entries $a_{i,j}$, for $k < i \leq m - 1$ and $1 \leq j \leq n$, arbitrarily from the set $\{1, 2, \ldots, p\}$. There are $p^{(m-k-1) \cdot n}$ ways to do so. Then the entries $a_{m,1}, a_{m,2}, \ldots, a_{m,\ell}$ are uniquely determined by the congruence relations

$$\sum_{i=1}^{m} a_{i,j} \equiv 0 \pmod{p} \quad \text{for all } 1 \leq j \leq \ell. \tag{$\dagger'$}$$

Finally, we can choose the entries $a_{m,j}$, with $\ell + 1 \leq j \leq n$, arbitrarily from $\{1, 2, \ldots, p\}$. There are $p^{n-\ell}$ ways to do so. It follows that there are

$$p^{k \cdot (n-1)} \cdot p^{(m-k-1) \cdot n} \cdot p^{n-\ell} = p^{mn-k-\ell}$$

such matrices; that is,

$$\left| M \cap \bigcap_{i=1}^{k} A_i \cap \bigcap_{i=1}^{\ell} B_j \right| = p^{mn-k-\ell} \quad \text{for } 0 \leq k < m \text{ and } 0 \leq \ell \leq n.$$

Similarly, we can show that

$$\left| M \cap \bigcap_{i=1}^{k} A_i \cap \bigcap_{i=1}^{\ell} B_j \right| = p^{mn-k-\ell} \quad \text{for } 0 \leq k \leq m \text{ and } 0 \leq \ell < n.$$

Combining the above yields

$$\left| M \cap \bigcap_{i=1}^{k} A_i \cap \bigcap_{i=1}^{\ell} B_j \right| = p^{mn-k-\ell}$$

for $0 \leq k \leq m$, $0 \leq \ell \leq n$, and $k + \ell < m + n$.

In the case $k = m$, $l = n$, if a matrix $A = (a_{i,j})$ belongs to $M \cap \bigcap_{i=1}^m A_i \cap \bigcap_{j=1}^n B_j$, then we can first pick the entries $a_{i,j}$, for $1 \le i \le m - 1$ and $1 \le j \le n - 1$, arbitrarily from $\{1, 2, \ldots, p\}$. There are $p^{(m-1)\cdot(n-1)} = p^{mn-m-n+1}$ ways to do so. Then all other entries are uniquely determined by the congruence relations (†) and (†′). We need to be a bit careful about $a_{m,n}$, because it appears twice in the above congruence relations. Indeed, $a_{m,n}$ is well-defined, because modulo p,

$$
\begin{aligned}
a_{m,n} &\equiv -(a_{m,1} + a_{m,2} + \cdots + a_{m,n-1}) \\
&\equiv -\left[-\sum_{i=1}^{m-1} a_{i,1} - \sum_{i=1}^{m-1} a_{i,2} - \cdots - \sum_{i=1}^{m-1} a_{i,n-1} \right] \\
&\equiv \sum_{j=1}^{n-1} \sum_{i=1}^{m-1} a_{i,j} = \sum_{i=1}^{m-1} \sum_{j=1}^{n-1} a_{i,j} \\
&\equiv -\left[-\sum_{j=1}^{n-1} a_{1,j} - \sum_{j=1}^{n-1} a_{2,j} - \cdots - \sum_{j=1}^{n-1} a_{m-1,j} \right] \\
&\equiv -(a_{1,n} + a_{2,n} + \cdots + a_{m-1,n}).
\end{aligned}
$$

Combining the above, we have

$$
\begin{aligned}
N &= \sum_{\substack{0 \le k \le m \\ 0 \le \ell \le n \\ k \ne m, \ell \ne n}} \left[\binom{m}{k} \binom{n}{\ell} (-1)^{k+\ell} p^{mn-k-\ell} \right] + (-1)^{m+n} p^{mn-m-n+1} \\
&= \sum_{\substack{0 \le k \le m \\ 0 \le \ell \le n}} \left[\binom{m}{k} \binom{n}{\ell} (-1)^{k+\ell} p^{mn-k-\ell} \right] \\
&\quad - (-1)^{m+n} p^{mn-m-n} + (-1)^{m+n} p^{mn-m-n+1} \\
&= \sum_{s=0}^{m+n} \sum_{\substack{k+\ell=s \\ k,\ell \ge 0}} \left[\binom{m}{k} \binom{n}{\ell} (-1)^s p^{mn-s} \right] + (-1)^{m+n} p^{mn-m-n}(p-1) \\
&= p^{mn-m-n} \left[\sum_{s=0}^{m+n} \left[\binom{m+n}{s} (-1)^s p^{m+n-s} \right] + (-1)^{m+n}(p-1) \right] \\
&= p^{mn-m-n} \left[(p-1)^{m+n} + (-1)^{m+n}(p-1) \right],
\end{aligned}
$$

by the **Vandermonde Identity** (Example 3.10) and then by Theorem 3.1.   ■

Note that in the solution of Example 6.14, we used the sum transformations

$$\sum_{i=1}^{m}\sum_{j=1}^{n} = \sum_{j=1}^{n}\sum_{i=1}^{m} \quad \text{and} \quad \sum_{\substack{0\leq k\leq m \\ 0\leq \ell\leq n}} = \sum_{s=0}^{m+n}\sum_{\substack{k+\ell=s \\ k,\ell\geq 0}}.$$

The idea of switching the order of summation is fundamental to Calculating in Two Ways, which is the central topic of the next Chapter.

Exercises 6

6.1. [AMC12 2001] How many positive integers not exceeding 2001 are multiples of 3 or 4 but not 5?

6.2. Adrian is to arrange three identical red marbles, four identical green marbles, and five identical blue marbles in such a way that all marbles of the same color are not together. In how many ways can Adrian finish this task?

6.3. [Canada 1983] Let n be a positive integer, and let $S = \{1, 2, \ldots, n\}$. Show that the number of permutations of S with no fixed points and the number of permutations of S with exactly one fixed point differ by 1.

6.4. Prove Example 6.5a.

6.5. Prove Theorem 6.3.

6.6. Let n be a positive integer. Develop a general formula for the number of positive integer divisors of n^2 that are smaller than n but do not divide n.

6.7. For positive integers $x_1, x_2, \ldots, x_n$, let

$$[x_1, x_2, \ldots, x_n] \quad \text{and} \quad (x_1, x_2, \ldots, x_n)$$

denote their least common multiple and greatest common divisor, respectively. Prove that

$$\frac{[a, b, c]^2}{[a, b][b, c][c, a]} = \frac{(a, b, c)^2}{(a, b)(b, c)(c, a)}$$

for positive integers a, b, and c.

6.8. The director of student activities in a boarding school wants to distribute 61 concert tickets to three dorms in such a way that no dorm gets more tickets than the sum of the numbers of tickets the other two dorms get. In how many ways can this be done?

6.9. Let $S = \{1, 2, \ldots, n\}$ and let T be the set of all subsets of S (including the empty set and S itself). One tries to choose three (not necessarily distinct) sets from the set T such that either two of the chosen sets are subsets of the third set or one of the chosen sets is a subset of both of the other two sets. In how many ways can this be done?

6.10. Let S be a finite set, and let k be a positive integer. Determine the number of ordered k-tuples $(S_1, S_2, \ldots, S_k)$ of subsets of S such that $\bigcap_{i=1}^{k} S_i = \emptyset$.

6.11. Let $a_1, a_2, \ldots, a_n$ be positive integers, and let $[a_1, a_2, \ldots, a_n]$ denote their least common multiple. Prove that

$$[a_1, a_2, \ldots, a_n] \geq \frac{a_1 a_2 \cdots a_n}{\prod_{1 \leq i < j \leq n} \gcd(a_i, a_j)},$$

where $\gcd(x, y)$ denote the greatest common divisor of integers x and y.

6.12. For each positive real number x, let $S(x) = \{\lfloor kx \rfloor \mid k = 1, 2, \ldots\}$. Let $x_1, x_2,$ and x_3 be three real numbers greater than 1 such that $\frac{1}{x_1} + \frac{1}{x_2} + \frac{1}{x_3} > 1$. Prove that there exist i and j, with $1 \leq i < j \leq 3$, such that $S(x_i) \cap S(x_j)$ has infinitely many elements.

7
Calculating in Two Ways: Fubini's Principle

So far, most of the problems that we have presented, regardless of the level of difficulty or complexity, could have been solved with direct calculation, that is, each problem could be solved by counting the number of objects (maybe via bijections or recursions). In this Chapter, in order to find x, the number of objects of type A, we will find y, the number of objects of type B, in order to set up equations involving x, after which we will solve for x. In other words, the problems that we analyzed in the earlier chapters are Arithmetic problems, but now we will be doing Algebraic problems by setting up equations and solving them.

Example 7.1. A 15×15 square is tiled with unit squares. Each vertex is colored either red or blue. There are 133 red points. Two of those red points are corners of the original square, and another 32 red points are on the sides. The sides of the unit squares are colored according to the following rule: If both endpoints are red, then it is colored red; if the points are both blue, then it is colored blue; if one point is red and the other is blue, then it is colored yellow. Suppose that there are 196 yellow sides. How many blue segments are there?

Solution: There are 15 sides of unit squares in each row, and there are 16 rows. Hence there are $15 \cdot 16$ horizontal sides of unit squares. Similarly, there are $15 \cdot 16$ vertical sides of unit squares. Thus, there are a total of $30 \cdot 16 = 480$ sides. Because we are discussing Calculating in Two Ways, let us count this number in another way. There are $15^2 = 225$ unit squares. Each unit square has four sides. There are 60 unit sides on the sides of the original square. All of the sides inside the original square are counted twice, because each of them is a common edge to two unit squares. Thus, there are $(225 \cdot 4 + 60)/2 = 480$ sides.

There are $480 - 196 = 284$ sides that are either red or blue. Assume that there are r red sides. Then there are $284 - r$ blue sides.

Next we count the number of appearances of red vertices as endpoints of unit sides in two ways. Let $|S|$ denote this number.

There are two such appearances for each red side, one such appearance for each yellow side, and zero appearances for each blue side. Hence

$$|S| = 2r + 196.$$

On the other hand, each red vertex on the corner of the original square appears twice, each red vertex on the sides of the original square appears three times, and each red vertex in the interior of the original square appears four times. Thus,

$$|S| = 2 \cdot 2 + 32 \cdot 3 + (133 - 2 - 32) \cdot 4 = 496.$$

It follows that $2r + 196 = 496$ and $r = 150$. Therefore, there are $284 - 150 = 134$ blue sides. ∎

Let m and n be positive integers, and let $A = \{a_1, a_2, \ldots, a_m\}$ and $B = \{b_1, b_2, \ldots, b_n\}$ be two finite sets. Then the **direct product** of A and B is defined as

$$A \times B = \{(a,b) \mid a \in A, \quad b \in B\}.$$

Let S be a subset of $A \times B$. For $a_i \in A$, let $S(a_i, *) = \{(a_i, b) \in S\}$. We define $S(*, b_i)$ analogously.

Theorem 7.1. *Let m and n be positive integers, and let $A = \{a_1, a_2, \ldots, a_m\}$ and $B = \{b_1, b_2, \ldots, b_n\}$ be two finite sets. If S is a subset of $A \times B$, then*

$$|S| = \sum_{j=1}^{n} |S(*, b_j)| = \sum_{i=1}^{m} |S(a_i, *)|.$$

This principle is called **Fubini's Principle**. It parallels a technique that is used in double integration.

Proof: It might be easier to interpret this principle as adding up the entries of a matrix. It is easy to see that this sum is equal to the sum of the row sums as well as to the sum of the column sums. We consider an $m \times n$ matrix $\mathbf{M} = x_{(i,j)}$ such that

$$x_{i,j} = \begin{cases} 1 & \text{if } (a_i, b_j) \text{ is in } S, \\ 0 & \text{otherwise.} \end{cases}$$

Then $|S(*, b_j)|$ is the sum of the entries in the jth column. Similarly, $|S(a_i, *)|$ is the sum of the entries in the ith row. Therefore, both sums, $\sum_{j=1}^{n} |S(*, b_j)|$ and $\sum_{i=1}^{m} |S(a_i, *)|$, evaluate the sum of all of the entries in matrix $\mathbf{M}$, from which the result follows. ∎

The set-theoretic language used in Theorem 7.1 might seem abstract. What sets play the parts of A, B, and S in Example 7.1? We can let A be the set of vertices of unit squares, and let B be the set of sides. Then S is the set of pairs (v, s) where v is red and v is on s.

Example 7.2. Let n be a positive integer, and let $(a_1, a_2, \ldots, a_n)$ be a permutation of $(1, 2, \ldots, n)$. For $1 \le k \le n$, let

$$F_k = \{a_i \mid a_i < a_k, \quad i > k\} \quad \text{and} \quad G_k = \{a_i \mid a_i > a_k, \quad i < k\}.$$

Prove that $\sum_{k=1}^{n} |F_k| = \sum_{k=1}^{n} |G_k|$.

Solution: Let $A = B = \{a_1, a_2, \ldots, a_n\}$, and let

$$S = \{(a_i, a_j) \mid a_i < a_j, \quad i > j\}.$$

Then $S(*, a_j) = F_j$ and $S(a_i, *) = G_i$. The desired result follows directly from Theorem 7.1. ∎

Most of the time, it is difficult to figure out what objects need to be calculated in different ways. There are no clear rules about this. On the other hand, the problems themselves provide valuable information.

Example 7.3. There are 12 students in Mr. Fat's combinatorics class. At the beginning of each week, Mr. Fat assigns a project to his students. The students form six groups. Each group works on the project independently and submits the work at the end of the week. Each week, the students can form the groups as they wish. Prove that, regardless of the way the students choose their partners, there are always two students such that there are at least five other students who have all worked with both of them or have worked with neither of them. (A student cannot work by himself.)

Solution: This problem investigates the working partnership between a student and a pair of students.   Thus, we let $A = \{s_1, s_2, \ldots, s_{12}\}$ denote the set of all students, and let $B = \{(s_i, s_j) \mid 1 \leq i < j \leq 12\}$ denote the set of all pairs of students.   Then $|B| = \binom{12}{2} = 66$. We say that s_i and (s_j, s_k) are *connected* if $s_i, s_j,$ and s_k are distinct and s_i has worked with exactly one of s_j and s_k. Let

$$S = \{[s_i, (s_j, s_k)] \mid s_i \text{ and } (s_j, s_k) \text{ are connected}\}.$$

Now we prove the desired result. We approach indirectly by assuming that the statement is not true, that is, that at some time, every pair (s_j, s_k) is connected to at least six students, or $|S(*, (s_j, s_k))| \geq 6$. Consequently, we have

$$|S| = \sum_{(s_j, s_k) \in B} |S(*, (s_j, s_k))| \geq 6|B| = 396.$$

On the other hand, if s_i has worked with d partners, then s_i is connected to $d(11 - d)$ pairs of students (because we have d ways to choose one of his previous partners and $11 - d$ ways to choose a student who has not worked with him). For integers $0 \leq d \leq 11$, the maximum value of $d(11 - d)$ is 30, obtained when $d = 5$ or 6. Hence $|S(s_i, *)| \leq 30$, implying that

$$|S| = \sum_{s_i \in A} |S(s_i, *)| \leq 30|A| = 360.$$

We obtain $396 \leq |S| \leq 360$, which is impossible.   Thus, our assumption was wrong and the problem statement is true.   ∎

Interested readers might want to generalize the last result to n students.

Example 7.4. Let X be a finite set with $|X| = n$, and let $A_1, A_2, \cdots, A_m$ be three-element subsets of X such that $|A_i \cap A_j| \leq 1$ for all $i \neq j$. Show that there exists a subset A of X with at least $\lfloor \sqrt{2n} \rfloor$ elements containing none of the A_i.

Solution: Let A be a subset of X containing none of the A_i, with the maximum number of elements. Let $k = |A|$. The difficulty in this problem is how to use the maximality of A. We count the number of elements in $X \backslash A$ in different ways. Clearly, this number is $n - k$.

Let x be an element in X that is not in A. By the maximality of A, $A \cup \{x\}$ does not satisfy the conditions of the problem; that is, there exists $i(x) \in \{1, \ldots, m\}$ such that $A_{i(x)} \subseteq A \cup \{x\}$. Therefore, x is in $A_{i(x)}$, and the set $A_{i(x)} \backslash \{x\}$ is a subset of A. Hence, the set $L_x = A \cap A_{i(x)}$ must have 2 elements. Because $|A_i \cap A_j| \leq 1$ for $i \neq j$, the sets L_x must all be distinct.

Thus we have defined an injective map $f(x) = L_x$ from $X \backslash A$ to the set of two-element subsets of A. From this we conclude that

$$n - k \leq \binom{k}{2} = \frac{k^2 - k}{2},$$

or $k^2 + k \geq 2n$, so $k \geq \lfloor \sqrt{2n} \rfloor$ (note that $(\lfloor \sqrt{2n} \rfloor - 1)^2 + (\lfloor \sqrt{2n} \rfloor - 1) \leq \sqrt{2n}(\sqrt{2n} - 1) < 2n$). ∎

Example 7.5. [IMO 2002 Short–listed] Let n be a positive integer. A sequence of n (not necessarily distinct) positive integers is called *full* if it satisfies the following conditions: For each positive integer $k \geq 2$, if the number k appears in the sequence, then so does the number $k-1$; moreover, the first occurrence of $k-1$ comes before the last occurrence of k. For each n, how many full sequences are there?

The following solution belongs to Federico Ardila, leader of the Colombian IMO team. It uses a bijection, but it is written in the language of Calculating in Two Ways.

Solution: We use an ordered 4-tuple of positive integers to classify full sequences, more precisely, (n, k, a, b) denotes the set of all full n-sequences where the largest term is k, and it appears for the first time in the ath term and the last time in the bth term. Let $f(n, k, a, b)$ denote the number of elements in (n, k, a, b). In this problem, we want to find

$$f(n) = \sum_{\substack{1 \leq k \leq n \\ 1 \leq a \leq b \leq n}} f(n, k, a, b).$$

Consider an element in (n, k, a, b), and remove its first k. (This is the bijection!) We consider two cases:

- If $a < b$, then we obtain an element of $(n-1, k, x, b-1)$ for some $a \leq x \leq b - 1$. Thus

$$f(n, k, a, b) = \sum_{a \leq x \leq b-1} f(n - 1, k, x, b - 1).$$

- If $a = b$, then we have only one k in the original sequence. Thus there are no k's in the new sequence. We obtain an element in $(n-1, k-1, x, y)$ for some $x \leq a-1$ (because $k-1$ first appeared before the last occurrence of k) and $x \leq y$. Then

$$f(n, k, a, a) = \sum_{\substack{1 \leq x \leq a-1 \\ x \leq y}} f(n-1, k-1, x, y).$$

Now we consider the sum

$$f(n) = \sum_{\substack{1 \leq k \leq n \\ 1 \leq a \leq b \leq n}} f(n, k, a, b)$$

$$= \sum_{1 \leq k \leq n} \sum_{1 \leq a < b \leq n} \sum_{a \leq x \leq b-1} f(n-1, k, x, b-1)$$

$$+ \sum_{1 \leq k \leq n} \sum_{1 \leq a \leq n} \sum_{\substack{1 \leq x \leq a-1 \\ x \leq y}} f(n-1, k-1, x, y).$$

For $1 \leq k \leq n-1$ and $1 \leq x \leq y \leq n-1$, how many times does $f(n-1, k, x, y)$ appear as a summand on the right-hand side of the above equation?

- It appears as $f(n-1, k, x, (y+1)-1)$ in

$$\sum_{1 \leq a < b \leq n} \sum_{a \leq x \leq b-1} f(n-1, k, x, b-1)$$

as a summand for each $1 \leq a \leq x$. Hence it appears x times as a summand.

- It appears as $f(n-1, (k+1)-1, x, y)$ in

$$\sum_{1 \leq a \leq n} \sum_{\substack{1 \leq x \leq a-1 \\ x \leq y}} f(n-1, k-1, x, y)$$

as a summand for each $x \leq a-1$, or $x+1 \leq a \leq n$. Hence it appears $n-x$ times as a summand.

Hence each $f(n-1, k, x, y)$ appears $x + (n-x) = n$ times as a summand, and

$$f(n) = \sum_{\substack{1 \leq k \leq n-1 \\ 1 \leq x \leq y \leq n-1}} n f(n-1, k, x, y) = n f(n-1).$$

Because $f(1) = 1$, we have $f(n) = n!$.   ■

Various solutions are possible for the above problem. Most of them use bijections. We provide a hint for one of the better bijections known. It is not difficult to see that there are $n!$ ways to place n non-attacking rooks on an $n \times n$ chessboard (Exercise 2.7). For each of such placements, we can find a unique full sequence corresponding to it. For example, for $n = 6$, we can map the following placements of Figure 7.1

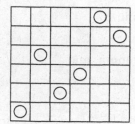 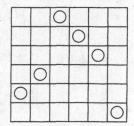

Figure 7.1.

to sequences $(3, 2, 3, 3, 1, 2)$ and $(4, 3, 1, 2, 3, 5)$, respectively. Can you define this mapping and prove it is indeed a bijection?

A classic example of applying the technique of Calculating in Two ways is the proof by Lubell, Meshalkin, and Yamamoto of a theorem of E. Sperner, from 1928.

Theorem 7.2. *[Sperner] Let S be a set with $|S| = n$. Assume that $S_1, S_2, \ldots, S_m$ are subsets of S such that $S_i \not\subseteq S_j$ for $i \neq j$. Then*

$$m \leq \binom{n}{\lfloor \frac{n}{2} \rfloor}.$$

Proof: An ordered n-tuple of sets $(T_1, T_2, \ldots, T_n)$ is called a *maximal chain* of S if

$$\emptyset \neq T_1 \subset T_2 \subset T_3 \subset \cdots \subset T_n = S.$$

We call T_i a *link* of the chain. Because T_i is a proper subset of T_{i+1}, the set T_i has i elements. There are n ways to define T_1. After T_1 is defined, there are $n - 1$ ways to define T_2, and so on. Hence there are $n!$ maximal chains of S if $|S| = n$.

For a set S_i, let us consider the maximal chains in S that contain S_i as a link. Assume that $(T_1, T_2, \ldots, T_n)$ is such a chain. If $|S_i| = k_i$, then $T_{k_i} = S_i$. Note that $(T_1, T_2, \ldots, T_{k_i})$ is a maximal chain of S_i, so there are $k_i!$ such chains. Because there are $n - k_i$ elements in $S - S_i$, there are $n - k_i$ ways to decide T_{k_i+1}, and so on. There are $(n - k_i)!$

ways to choose $T_{k_i+1}, \ldots, T_n$. Thus there are $k_i!(n-k_i)!$ maximal chains of S that contain S_i as a link.

Let A be the set of all maximal chains and let $B = \{S_1, S_2, \ldots, S_m\}$. Let

$$S = \{(C_i, S_j) \mid C_i \in A, \ S_j \in B, \ S_j \text{ is a link of } C_i\}.$$

As in the proof of Theorem 7.1, we define a matrix $\mathbf{M} = (x_{i,j})$ with $n!$ rows and m columns such that

$$x_{i,j} = \begin{cases} 1 & \text{if } (C_i, S_j) \text{ is in } S, \\ 0 & \text{otherwise.} \end{cases}$$

Thus there are $k_j!(n-k_j)!$ 1's in the jth column, and

$$|S| = \sum_{j=1}^{m} k_j!(n-k_j)!.$$

On the other hand, because S_i is not a subset of S_j when $i \neq j$, S_i and S_j cannot be links of the same maximal chain of S. Thus the sum of the elements in each row of $\mathbf{M}$ is at most 1. Hence the sum of all of the row sums is at most $n!$, or

$$|S| \leq n!.$$

It follows that

$$\sum_{j=1}^{m} k_j!(n-k_j)! \leq n!,$$

or

$$\sum_{j=1}^{m} \frac{1}{\binom{n}{k_j}} \leq 1.$$

By Theorem 3.2 (c), $\binom{n}{k_j} \leq \binom{n}{\lfloor \frac{n}{2} \rfloor}$, so

$$m \frac{1}{\binom{n}{\lfloor \frac{n}{2} \rfloor}} \leq \sum_{j=1}^{m} \frac{1}{\binom{n}{k_j}} \leq 1,$$

as desired. Note that $m = \binom{n}{\lfloor \frac{n}{2} \rfloor}$ is obtained when B consists of all $\lfloor \frac{n}{2} \rfloor$-element subsets of S. ∎

Next we will present four geometric examples.

Example 7.6. [IMO 2000 Short–listed] Let $n \geq 4$ be a fixed positive integer. Let $S = \{P_1, P_2, \ldots, P_n\}$ be a set of n points in the plane lie on the same circle). By circle $P_i P_j P_k$ we mean the circumcircle of the triangle $P_i P_j P_k$. Let a_t, $1 \le t \le n$, denote the number of circles $P_i P_j P_k$ that contain P_t in their interiors, and let

$$m(S) = a_1 + a_2 + \cdots + a_n.$$

Prove that there exists a positive integer $f(n)$ depending only on n such that the points of S are the vertices of a convex polygon if and only if $m(S) = f(n)$.

Solution: If $ABCD$ is a convex, noncyclic quadrilateral, then $\angle A + \angle C \ne \angle B + \angle D$. Because the sum of these four angles is $360°$, we may assume without loss of generality that $\angle A + \angle C > 180° > \angle B + \angle D$. Then B and D are outside of circles ACD and ABC, respectively, and A and C are inside of circles BCD and BAD, respectively. If $ABCD$ is a concave quadrilateral, we assume without loss of generality that $\angle A > 180°$. Then A is inside of circle BCD, and B, C, and D are outside of circles ACD, ABD, and ABC, respectively. For each set $\{P_i, P_j, P_k, P_\ell\}$, there are four possible pairs that can contribute a 1 to the quantity $m(S)$, namely $(P_i, P_j P_k P_\ell)$, $(P_j, P_k P_\ell P_i)$, $(P_k, P_\ell P_i P_j)$, and $(P_\ell, P_i P_j P_k)$. If the four points form a convex polygon, then exactly two of these pairs contribute a 1 to $m(S)$. Otherwise, only one of them does.

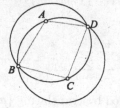 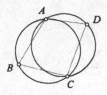

Figure 7.2.

Let $a(S)$ and $b(S)$ denote the numbers of convex and concave quadrilaterals determined by the points of S, respectively. We have

$$a(S) + b(S) = \binom{n}{4} \quad \text{and} \quad 2a(S) + b(S) = m(S).$$

Consequently, $m(S) = \binom{n}{4} + a(S)$. We claim that $f(n) = 2\binom{n}{4}$ has the desired property. Indeed, if the points of S form a convex polygon, then every quadrilateral determined by them is convex, so $a(S) = \binom{n}{4}$ and $m(S) = f(n)$. Conversely, if $m(S) = f(n)$, then $a(S) = \binom{n}{4}$, so that every quadrilateral determined by four points of S is convex. It follows easily that the points of S form a convex polygon.   ∎

Example 7.7. The coordinate plane is tiled with unit squares in such a way that $(0,0)$, $(1,0)$, $(1,1)$, and $(0,1)$ are the vertices of a unit square. For a set S of unit squares, another set of unit squares is a *translation* of S if it can be obtained by translating all of the squares of S by a common vector. (We consider S to be a translation of itself, because it can be obtained by translating S by $[0,0]$.) A real number is written in each unit square. For any set S of unit squares, the *value* of S is defined as the sum of all of the numbers written in the squares of S. Assume that there is a finite set A of unit squares such that the values of A and all of its translations are positive. Let B be another finite set of unit squares. Show that there is a translation of B with a positive value.

Solution: Let $\{x, y\}$ denote the unit square with (x, y) as its bottom left corner, and let $r(x, y)$ denote the number written in $\{x, y\}$. Let

$$A = \{\{a_1, b_1\}, \{a_2, b_2\}, \ldots, \{a_n, b_n\}\}$$

and

$$B = \{\{u_1, v_1\}, \{u_2, v_2\}, \ldots, \{u_m, v_m\}\}.$$

We consider the sum

$$s = \sum_{i=1}^{m} \sum_{j=1}^{n} r(a_i + u_j, b_i + v_j) = \sum_{j=1}^{m} \sum_{i=1}^{n} r(a_i + u_j, b_i + v_j).$$

For each j, $\{a_i + u_j, b_i + v_j\}$ is obtained by translating $\{a_i, b_i\}$ by the vector $[u_j, v_j]$. By the given condition, the value of the translation of A by $[u_j, v_j]$ is positive, that is,

$$\sum_{i=1}^{n} r(a_i + u_j, b_i + v_j) > 0 \quad \text{for all } j,$$

implying that $s = \sum_{j=1}^{m} \sum_{i=1}^{n} r(a_i + u_j, b_i + v_j) > 0$. Therefore, for some i,

$$\sum_{j=1}^{m} r(a_i + u_j, b_i + v_j) > 0,$$

because otherwise, $s = \sum_{i=1}^{n} \sum_{j=1}^{m} r(a_i + u_j, b_i + v_j) \leq 0$. Consequently, the value of the translation of B by $[a_i, b_i]$ is positive, as desired. ∎

The key idea of the proof is to interpret the sum S in two ways: It is obtained by translating A by vectors in B as well as by translating B by vectors in A. Reader is encouraged to decompose the region S in two ways in Figure 7.3. Note that some squares are counted more than once in S. But each square is counted the same amount of times in two ways.

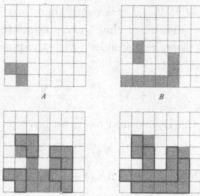

Figure 7.3.

Example 7.8. Let n be an integer with $n \geq 3$. Suppose that $P_1 P_2 \ldots P_n$ is a regular n-sided polygon. Determine the number of incongruent triangles of the form $P_i P_j P_k$, where i, j, and k are distinct integers between 1 and n, inclusive.

This problem can be solved using case analysis. Here is a more direct approach.

Solution: It is clear that these points are evenly distributed around a circle. By symmetry, it suffices to consider incongruent triangles that have P_1 as a vertex. Let N_1 be the number of incongruent equilateral triangles, N_2 the number of incongruent, nonequilateral isosceles triangles, and N_3 the number of incongruent scalene triangles. Thus we need to calculate $N = N_1 + N_2 + N_3$.

There are $\binom{n-1}{2} = \frac{(n-1)(n-2)}{2}$ triangles with P_1 as a vertex. Let $S = \{P_1 P_i P_j \mid 2 \leq i < j \leq n\}$. Then $|S| = \frac{(n-1)(n-2)}{2}$. In S, each scalene triangles appears six times, because there are 3! ways to assign the three lengths to the sides $P_1 P_i$, $P_i P_j$, and $P_j P_1$; each nonequilateral isosceles triangles appears three times, because there are three ways to choose $\overline{P_1P_i}$, $\overline{P_iP_j}$, $\overline{P_jP_1}$ as the base of the triangle; and the equilateral triangle, if there is one, appears once. Therefore,

$$\frac{(n-1)(n-2)}{2} = N_1 + 3N_2 + 6N_3. \tag{$*$}$$

It is also clear that there is at most one equilateral triangle with P_1 as a vertex. Hence $N_1 = 1$ or $N_1 = 0$, and $N_1 = 1 - p$ for some $p = 0$ or 1. To calculate N_2, we can always rotate the triangle so that P_1 is the vertex of the isosceles triangle. Hence we consider triangles $P_1P_iP_j$ with $P_1P_i = P_1P_j$. If n is even, there are $\frac{n-2}{2}$ such isosceles triangles (with the possibility that one is equilateral). If n is odd, there are $\frac{n-1}{2}$ such isosceles triangles (with the possibility that one is equilateral). Hence

$$N_1 + N_2 = \frac{n-2+q}{2}, \tag{$**$}$$

where $q = 0$ or $q = 1$. By $(*)$ and $(**)$, we have

$$\begin{aligned}
12N &= 12(N_1 + N_2 + N_3) \\
&= 2(N_1 + 3N_2 + 6N_3) + 6(N_1 + N_2) + 4N_1 \\
&= (n-1)(n-2) + 3(n-2+q) + 4(1-p) \\
&= n^2 + 3q - 4p,
\end{aligned}$$

where $p, q \in \{0, 1\}$. Since $-4 \le 3q - 4p \le 3$, N is the integer closest to $\frac{n^2}{12}$; that is, $N = \left\lfloor \frac{n^2}{12} + \frac{1}{2} \right\rfloor$. ■

Example 7.9. Let n and k be positive integers satisfying the following property: There exists a set T of n points in the plane such that

(i) no three are collinear;

(ii) for any point P in T, there are at least k points in T that are equidistant from P.

Prove that

$$k < \frac{1}{2} + \sqrt{2n}.$$

Solution: Let $A = T = \{P_1, P_2, \ldots, P_n\}$, and let $B = \{\ell_{i,j} \mid 1 \le i < j \le n\}$, where $\ell_{i,j}$ denotes the perpendicular bisector of $\overline{P_iP_j}$.

Then $|B| = \binom{n}{2}$. Let

$$S = \{(P_i, \ell_{j,k}) \mid P_i \text{ is on } \ell_{j,k}\}.$$

Since no three points are on a line, we have $|S(*, \ell_{j,k})| \leq 2$. Thus,

$$S = \sum_{\ell_{j,k} \in B} |S(*, \ell_{j,k})| \leq 2|B| = n^2 - n.$$

On the other hand, because P_i is equidistant to at least k other points, P_i is on the perpendicular bisector of any two of those k points. Hence $|S(P_i, *)| \geq \binom{k}{2}$, so

$$S = \sum_{P_i \in A} |S(P_i, *)| \geq n\binom{k}{2} = \frac{n(k^2 - k)}{2}.$$

Combining the above gives

$$k^2 - k - 2(n-1) \leq 0.$$

Solving this inequality yields

$$\frac{1}{2} - \sqrt{2n - \frac{7}{4}} \leq k \leq \frac{1}{2} + \sqrt{2n - \frac{7}{4}},$$

from which the desired result follows.   ∎

Example 7.10. [China 2000, HongBin Yu] Given n collinear points, consider the distances between the points. Suppose each distance appears at most twice. Prove that there are at least $\lfloor n/2 \rfloor$ distances that appear once each.

This problem seems rather harmless, but in reality, it is exactly the opposite. Indeed, no complete solutions were presented by the students on a MOSP practice test. All students tried to start with the *obvious* approach of using induction on n. Then they found out that it was very hard to reconstruct the conditions to apply the induction hypothesis. At first glance, it does not seem impossible to do so. However, whenever a gap in the argument is to be patched, a new "smaller gap" appears. In the end, people run out of time before they could patch up all of the gaps in the inductive approach. The bottom line is that we do not even know of a way to solve this problem by induction. The following solution was written by Po-Ru Loh, a three-time IMO medalist (two golds in 2002 and 2003 and one silver in 2000), with some hints from Zhongtao Wu (IMO gold medalist in 2000), who attended Professor Yu's lecture back in China.

Solution: Let x be the number of distances that appear once, and y the number of distances that appear twice. We would like to find a lower bound on the total number of distinct distances, $x + y$. To do this, we count the segments from left to right, according to their left end points. Denote the points by $P_1, P_2, \ldots, P_n$, from left to right. Then P_1 is the left end point of $n - 1$ segments with distinct lengths. Now we move on to P_2. This point is the left endpoint of $n - 2$ segments, but some may have lengths already counted by P_1. We claim that this can be the case for no more than one length. For if $P_1P_i = P_2P_j$ and $P_1P_k = P_2P_l$, then $P_1P_2 = P_iP_j = P_kP_l$, contradicting the assumption that each distance can appear at most twice. Hence no more than one length is repeated from P_1's count, and we have at least $n-3$ new lengths. Moving on to P_3, we can apply a similar analysis. There are $n - 3$ segments with left end point P_3, but one distance may already have been counted by P_1 and another by P_2. Hence we must lower our bound to $n-5$. We may continue in this fashion to arrive at a total lower bound of $(n-1)+(n-3)+(n-5)+\cdots$. If n is odd, this sum is equal to $\frac{n^2-1}{4}$, while if n is even, it is equal to $\frac{n^2}{4}$.

We know that the total number of segments, $x+2y$, is $\frac{n(n-1)}{2}$. From the above, we have $x + y \geq \lfloor \frac{n^2}{4} \rfloor$, so that $2x + 2y \geq \lfloor \frac{n^2}{2} \rfloor$. Therefore, $x \geq \lfloor \frac{n^2}{2} \rfloor - (\frac{n^2}{2} - \frac{n}{2}) = \lfloor \frac{n}{2} \rfloor$, as desired. ∎

Calculating in Two Ways is very useful in discussing partitions of integers. We recall the Young diagrams used in the solution to Example 4.14. The numbers of circles in the columns are parts of the partitions. Summing the numbers of circles in the rows can reveal interesting relations between the partitions.

Example 7.11. Let $p(n)$ denote the number of partitions of n, let $p(n,m)$ the number of partitions of n with length m, and $h(n,m)$ the number of partitions of n with height m. Prove that $p(n,m) = h(n,m)$.

Solution: If $(a_1, a_2, \ldots, a_m)$ is an m-part partition of n, then we can arrange n circles into m columns such that the jth column containing a_j circles. Then the first row contains exactly m circles. Thus, the number of circles in each row yields a partition of n with height m. In other words, we can define a matrix $\mathbf{M} = (x_{i,j})$ of a_m rows and m

columns as

$$x_{i,j} = \begin{cases} 1 & \text{if } i \le a_j, \\ 0 & \text{otherwise.} \end{cases}$$

The column sums of this matrix are $a_1, a_2, \ldots, a_m$. The row sums are $(m, \ldots)$. If we reverse the order of the rows, we obtain a partition of n with height m. It is easy to see that each partition of n with length m uniquely determines a matrix $\mathbf{M}$, and that each such matrix uniquely determines a partition of n with height m. Thus $p(n,m) = h(n,m)$, as desired. (For example, $n = 41$, $m = 6$, and the partition $\pi = (3, 5, 7, 8, 9, 9)$ are mapped as shown in Figure 7.3.

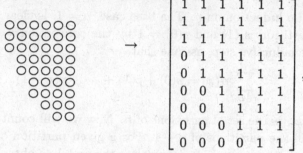

Figure 7.4.

It is then mapped to the partition $(2, 3, 4, 4, 5, 5, 6, 6, 6)$ with height 6.)   ∎

Example 7.12. If π is a partition of n, let $\alpha(\pi)$ denote the number of 1's in the parts of π, and let $\beta(\pi)$ denote the number of distinct parts of π. Prove that

$$\sum \alpha(\pi) = \sum \beta(\pi),$$

where the sum is taken over all partitions of n.

Solution: Let $A = \{1, 2, \ldots, n\}$, and let B denote the set of all partitions of n. As usual, let $p(n)$ denote the number of partitions of n, so that $|B| = p(n)$. We define

$$S = \{(a, \pi) \mid a \in A, \ \pi \in B, \ a \text{ is a part of } \pi\}.$$

Then $\beta(\pi) = |S(*, \pi)|$, and so by Theorem 7.1, we have

$$\sum_{\pi \in B} \beta(\pi) = \sum_{\pi \in B} |S(*, \pi)| = |S| = \sum_{a \in A} |S(a, *)|.$$

Let $\pi = (a_1, a_2, \ldots, a_m)$ be a partition of n. We note that if $a = a_k$ is a part of π, then $(a_1, \ldots, a_{k-1}, a_{k+1}, \ldots, a_m)$ is a partition of $n - a$. Hence $S|(a, *)| = p(n - a)$. (If $a = n$, there is only one such partition, namely, (n); thus we set $p(0) = 1$.) It follows that

$$\sum_{\pi \in B} \beta(\pi) = \sum_{a \in A} p(n - a) = \sum_{a=1}^{n} p(n - a) = \sum_{i=0}^{n-1} p(i).$$

It suffices to show that

$$\sum_{\pi \in B} \alpha(\pi) = p(0) + p(1) + p(2) + \cdots + p(n - 1).$$

We use induction on n. The base case, $n = 1$, is clear because the only partition is (1) and $p(0) = 1$ by the convention we set earlier. For the inductive step, assume that

$$\sum_{\pi \in B_n} \alpha(\pi) = p(0) + p(1) + \cdots + p(n - 1),$$

where B_n is the set of partitions of n. Now we will count the number of 1's in the partitions of $n + 1$. For a given partition of $n + 1$ that includes at least one 1, we may delete the first 1 to obtain a partition of n. Furthermore, $\sum_{\pi \in B_n} \alpha(\pi)$ counts the number of 1's in these partitions excluding the first 1's in the partitions. Clearly, the number of such partitions is $p(n)$, so that we must add on $p(n)$ to include those 1's. Thus

$$\sum_{\pi \in B_{n+1}} \alpha(\pi) = \sum_{\pi \in B_n} \alpha(\pi) + p(n) = p(0) + p(1) + \cdots + p(n - 1) + p(n),$$

and our induction is complete.   ∎

The next two examples require repeatedly calculating the numbers of different kinds of objects in different ways.

Example 7.13. [China 1997, by Zhusheng Zhang] Let n be an integer greater than six, and let X be an n-element set. Let $A_1, A_2, \ldots, A_m$ be distinct 5-element subsets of X. Prove that for

$$m > \frac{n(n - 1)(n - 2)(n - 3)(4n - 15)}{600},$$

there exist distinct indices $i_1, i_2, \ldots, i_6$ such that the union of the sets $A_{i_1}, A_{i_2}, \ldots, A_{i_6}$ contains exactly six elements.

Solution: Because each A_i has five elements and all A_i's are distinct, the union of any two of them contains at least six elements. We approach indirectly by assuming that for an integer $n > 6$ and some integer

$$m > \frac{n(n-1)(n-2)(n-3)(4n-15)}{600},$$

there are distinct 5-element subsets $A_1, A_2, \ldots, A_m$ such that the union of any six of them has more than six elements. Let $\mathcal{A} = \{A_1, A_2, \ldots, A_m\}$, and let $\mathcal{Q}$ denote the set of all four-element subsets of S contained in at least one of the elements in $\mathcal{A}$, that is,

$$\mathcal{Q} = \{Q \mid |Q| = 4, \ Q \subset A \text{ for some } A \in \mathcal{A}\}.$$

For an element A in $\mathcal{A}$, we call a four-element subset of A a *quartet*. Let $\mathcal{Q}_A$ denote the set of quartets of an element A in $\mathcal{A}$. Then $\mathcal{Q}$ is the set of all quartets. By this definition, for each Q in $\mathcal{Q}$ there is at least one element x in X such that $Q \cup \{x\} \in \mathcal{A}$. Let S_Q denote the union of all such x's, that is,

$$S_Q = \{x \mid x \in X, \ Q \cup \{x\} \in \mathcal{A}\}.$$

For an element A in $\mathcal{A}$, we now consider the direct product $\mathcal{Q}_A \times X$ and its subset

$$V_A = \{(Q, x) \mid Q \in \mathcal{Q}_A, \ x \in X, \ Q \cup \{x\} \in \mathcal{A}\}.$$

Then $|V_A(Q, *)| = |S_Q|$. By Theorem 7.1, we have

$$\sum_{Q \in \mathcal{Q}_A} |S_Q| = |V_A| = \sum_{x \in X} |V_A(*, x)| = \sum_{x \in X - A} |V_A(*, x)| + \sum_{x \in A} |V_A(*, x)|.$$

If $x \in A$, then $V_A(*, x) = \{(A - \{x\}, x)\}$, and $|V_A(*, x)| = 1$. If $x \in X - A$, then $V_A(*, x)$ can have no more than four elements. Otherwise, if $Q_1, Q_2, \ldots, Q_5$ are distinct quartets of $V_A(*, x)$, then the sets $A, A_{i_1}, A_{i_2}, \ldots, A_{i_5}$ are distinct elements of $\mathcal{A}$, where $A_{i_j} = Q_j \cup \{x\}$. (For $1 \le j < k \le 5$, A_{i_j} is different from A_{i_k} because Q_j and Q_k are distinct. The set A is different from A_{i_j} because $x \notin A$ and $x \in A_{i_j}$.) The union of these six sets is $A \cup \{x\}$, which has exactly six elements. This contradicts our assumption. Hence $|V_A(*, x)| \le 4$ for $x \in X - A$. Thus, we have

$$\sum_{Q \in \mathcal{Q}_A} |S_Q| = \sum_{x \in X - A} |V_A(*, x)| + \sum_{x \in A} |V(*, x)| \le 4(n-5) + 5 = 4n - 15,$$

and consequently,

$$\sum_{A\in\mathcal{A}}\sum_{Q\in\mathcal{Q}_A} |S_Q| \le m(4n-15).$$

Now we calculate the sum

$$\sum_{A\in\mathcal{A}}\sum_{Q\in\mathcal{Q}_A} |S_Q|$$

in different ways. Because each A in $\mathcal{A}$ is a 5-element set, we conclude that S_Q is a subset of $X-Q$, and that for $x\ne y$ in S_Q, $A_i = Q\cup\{x\} \ne Q\cup\{y\} = A_j$. In other words, there is an obvious bijection between S_Q and

$$T_Q = \{A \mid A\in\mathcal{A},\ Q\subset A \text{ or } Q\in\mathcal{Q}_A\}$$

for each quartet Q. Consequently, $|S_Q| = |T_Q|$. It follows that

$$\sum_{A\in\mathcal{A}}\sum_{Q\in\mathcal{Q}_A} |T_Q| = \sum_{A\in\mathcal{A}}\sum_{Q\in\mathcal{Q}_A} |S_Q| \le m(4n-15).$$

By the definition of T_Q, the summand $|T_Q|$ appears $|T_Q|$ times in the sum $\sum_{A\in\mathcal{A}}\sum_{Q\in\mathcal{Q}_A}|T_Q|$. Therefore, we have

$$m(4n-15) \ge \sum_{A\in\mathcal{A}}\sum_{Q\in\mathcal{Q}_A} |T_Q| = \sum_{Q\in\mathcal{Q}} |T_Q|^2.$$

By the **(Quadratic) Root Mean Square–Arithmetic Mean Inequality**, we have

$$\sum_{Q\in\mathcal{Q}} |T_Q|^2 \ge \frac{1}{|\mathcal{Q}|}\left[\sum_{Q\in\mathcal{Q}} |T_Q|\right]^2.$$

There are at most $\binom{n}{4}$ quartets, or $|\mathcal{Q}| \le \binom{n}{4}$. Putting the last three inequalities together yields

$$m(4n-15) \ge \frac{1}{\binom{n}{4}}\left[\sum_{Q\in\mathcal{Q}} |T_Q|\right]^2,$$

or

$$\frac{n(n-1)(n-2)(n-3)(4n-15)}{24} \ge \frac{1}{m}\left[\sum_{Q\in\mathcal{Q}} |T_Q|\right]^2. \quad (\dagger)$$

Consider the direct product $\mathcal{A}\times\mathcal{Q}$ and its subset

$$U = \{(A,Q) \mid A\in\mathcal{A}, Q\in\mathcal{Q},\ Q\subset A\}.$$

Then $U(*, Q) = T_Q$, and $T(A, *)$ denotes the set of all quartets in A, so $|T(A, *)| = 5$. By Theorem 7.1, we have

$$\sum_{Q \in \mathcal{Q}} |T_Q| = \sum_{Q \in \mathcal{Q}} |U(*, Q)| = |U| = \sum_{A \in \mathcal{A}} |T(A, *)| = 5m.$$

Therefore, inequality (†) becomes

$$\frac{n(n-1)(n-2)(n-3)(4n-15)}{24} \geq \frac{1}{m}(5m)^2,$$

or

$$\frac{n(n-1)(n-2)(n-3)(4n-15)}{600} \geq m,$$

which violates our assumption on m. Thus our assumption was wrong, and the problem statement is true. ∎

The previous problem was very difficult because we had to calculate

$$\sum_{A \in \mathcal{A}} \sum_{Q \in \mathcal{Q}_A} |S_Q|$$

in four different ways, and two of them were based on a bijection between S_Q and T_Q.

Example 7.14. Mr. Fat is baking 68 different cakes with different kinds of cake mix. Some of the kinds of cake mix are sweetened. Each cake is made from five different kinds of mix with at least one kind of sweetened mix among them. It is known that for every three kinds of mix there is exactly one cake containing them. Prove that there exists at least one very sweet cake – a cake made from at least four kinds of sweetened mix.

Solution: We approach indirectly by assuming that there are no very sweet cakes. Suppose there are n different kinds of cake mix, denoted by $s_1, s_2, \ldots, s_n$. Among them, we assume that $s_1, s_2, \ldots, s_m$ are the sweetened ones.

We call every three kinds of mix a *triplet*. We will count the total number of triplets in two ways. It is clear that there are $\binom{n}{3}$ triplets. On the other hand, each cake contributes $\binom{5}{3} = 10$ triplets. Because each triplet comes from exactly one cake, we have $68 \cdot 10 = \binom{n}{3}$, so $n = 17$.

Assume that s_i is used in c_i cakes. For each of those cakes, there are $\binom{4}{2} = 6$ triplets containing s_i. Hence there are $6c_i$ triplets containing

s_i. On the other hand, there are $\binom{16}{2} = 120$ triplets containing s_i. Hence $c_i = 20$; that is, each kind of mix is used in exactly 20 cakes.

Assume that s_i and s_j $(1 \leq i < j \leq 17)$ are used together in $c_{i,j}$ cakes. For each of those cakes, there are $\binom{3}{1} = 3$ triplets containing both s_i and s_j. Hence there are $3c_{i,j}$ triplets containing s_i and s_j. On the other hand, there are $\binom{15}{1} = 15$ triplets containing s_i and s_j. Thus $c_{i,j} = 5$; that is, each pair of mixes is used together in exactly 5 cakes.

Assume that s_i, s_j, s_k $(1 \leq i < j < k \leq 17)$ are used together in $c_{i,j,k}$ cakes. By the given condition in the problem, $c_{i,j,k} = 1$. Assume that s_i, s_j, s_k, s_ℓ are used together in $c_{i,j,k,\ell}$ cakes. Under our assumption, $c_{i,j,k,\ell} = 0$ if $1 \leq i < j < k < \ell \leq m$.

Now what should we count? It seems that we have exhausted all of the conditions given in the problem. Not so fast my friend! There is one condition still untouched, namely, that at least one kind of sweetened mix is used in each cake. We count the total number of cakes in two ways. Obviously, this number is equal to 68. By the Inclusion–Exclusion Principle, we have

$$68 = \sum_{i=1}^{m} c_i - \sum_{1 \leq i < j \leq m} c_{i,j} + \sum_{1 \leq i < j < k \leq m} c_{i,j,k}$$

$$= 20m - 5\binom{m}{2} + \binom{m}{3}$$

$$= \frac{1}{6}m(m^2 - 18m + 137),$$

or

$$68 \cdot 6 = m(m^2 - 18m + 137).$$

It suffices to show that there is no such $m \leq 17$ (because $m \leq n = 17$). We can check case by case, which is not that bad. Instead, we count the equation modulo 5. Then the last equation becomes

$$3 \equiv m(m^2 - 3m + 2) \equiv m(m - 1)(m - 2) \pmod 5.$$

By checking $m = 0, \pm 1, \pm 2$, we conclude that the last equation has no integer solutions in m. Thus, our assumption was wrong, and there is at least one very sweet cake. Sweet! ∎

Exercises 7

7.1. Two exercises on addition:

(a) What is the sum of all possible odd four-digit numbers that can be formed by using each of the digits 2, 4, 5, 6, and 8 exactly once in each number?

(b) [ARML 2003] Let x and y be positive integers such that

$$2^x 3^y = \prod_{i=1}^{59}\left(24^{\frac{1}{i+1}+\frac{1}{i+2}+\cdots+\frac{1}{60}}\right)^i.$$

Find $x + y$.

7.2. Sixty students who took the AIME at PEA in 2003. The possible scores on the AIME are integers between 0 to 15, inclusive. Let $a_1, a_2, \ldots, a_{60}$ denote the students' scores on the AIME. For $k = 0, 1, \ldots, 15$, let b_k denote the number of students with a score of at least k. Show that

$$a_1 + a_2 + \cdots + a_{60} = b_1 + b_2 + \cdots + b_{15}.$$

7.3. [AIME 1993] A particular convex polyhedron has 32 faces, each of which is either a triangle or a pentagon. T triangular faces and P pentagonal faces meet at each of its V vertices. What is the value of $100P + 10T + V$?

7.4. [Russia 1990] There are 30 senators in a senate. Each pair of senators, the two senators are either friends of each other or enemies of each other. Every senator has exactly six enemies. Every three senators form a committee. Find the total number of committees whose members are either all friends or all enemies of each other.

7.5. Let n be a positive integer. For a permutation π of the set $S = \{1, 2, \ldots, n\}$, define

$$f(\pi) = \sum_{k=1}^{n} |k - \pi(k)|.$$

Express $\sum f(\pi)$ is closed form, where the sum is taken over all permutations of S.

7.6. A set M has seven elements. Let $A_1, A_2, \ldots, A_7$ be proper subsets of M such that

(i) each pair of elements of M belongs to exactly one of the subsets;

(ii) each subset has at least three elements.

Prove that every two subsets have exactly one common element.

7.7. Let n be a positive integer. There are n pieces of candy in a single pile. At each step, Zachary is allowed to choose a pile containing more than one piece of candy (if there is such a pile), separate it into two piles, and then calculate the product of the numbers of pieces of candy in the resulting two piles. Zachary keeps applying this operation until he obtains n piles of candy. Prove that the sum of all of the products Zachary obtains is independent of his procedure.

7.8. [IMC 2002] Two hundred students participated in a mathematical contest. They had six problems to solve. It is known that each problem was correctly solved by at least 120 participants. Prove that there must be two participants such that each problem was solved by at least one of the two students.

7.9. Eleven teachers run a conference. Every hour, one or more teachers give a one-hour presentations, while all of the other teachers observe the presentations. (If one chooses to observe a presentation, then he has to observe it for the whole period.) Find the least amount of time during which it is possible for each teacher to observe all other presentations at least once.

7.10. All points in a 100×100 array are colored in one of four colors red, green, blue, and yellow, in such a way that there are 25 points of each color in each row and in any column. Prove that there are two rows and two columns such that their four intersection points are all different colors.

7.11. [IMO 1998] In a competition there are a contestants and b judges, where $b \geq 3$ is an odd integer. Each judge rates each contestant with *pass* or *fail*. Suppose k is a number such that for any two judges, their ratings coincide for at most k contestants. Prove that

$$\frac{k}{a} \geq \frac{b-1}{2b}.$$

7.12. Let n be a positive integer. Let $d(n)$ denote the number of partitions of n with distinct parts, and let $o(n)$ denote the number of partitions of n with all odd parts. Prove that $d(n) = o(n)$.

8
Generating Functions

Let $A = (a_0, a_1, a_2, \ldots)$ be a sequence. Then

$$P_A(x) = a_0 + a_1 x + a_2 x^2 + \ldots$$

is the **generating function** (of the first type) associated with A, that is, the coefficient of x^n in the polynomial $P_A(x)$ is equal to the value of the nth term of A.

It should not be a surprise that generating functions are very useful in proving combinatorial identities, because the sequence of binomial coefficients $A = \left(\binom{n}{0}, \binom{n}{1}, \ldots, \binom{n}{n}\right)$ is exactly the sequence of coefficients of $P_A(x) = (1 + x)^n$.

Example 8.1. [AMC12 2003] Objects A and B move simultaneously in the coordinate plane via a sequence of steps, each of length one. Object A starts at $(0,0)$, and each of its steps is either right or up, both equally likely; object B starts at $(5,7)$, and each of its steps is either left or down, both equally likely. What is the probability that the objects meet?

Solution: Because there are twelve steps between $(0,0)$ and $(5,7)$, A and B can meet only after they have each moved six steps. The possible meeting places are $P_0 = (0,6)$, $P_1 = (1,5)$, $P_2 = (2,4)$, $P_3 = (3,3)$, $P_4 = (4,2)$, and $P_5 = (5,1)$. Let a_i and b_i denote the number of paths to P_i from $(0,0)$ and $(5,7)$, respectively. Because A has to take i steps to the right and B has to take $i+1$ steps down, the number of ways in which A and B can meet at P_i is

$$a_i \cdot b_i = \binom{6}{i} \cdot \binom{6}{i+1}.$$

Note also that the objects cannot meet twice. Because A and B can each take 2^6 paths in six steps, the probability that they meet is

$$\frac{1}{2^{12}} \sum_{i=0}^{5} a_i b_i = \frac{1}{2^{12}} \sum_{i=0}^{5} \binom{6}{i} \cdot \binom{6}{i+1} = \frac{99}{512}.$$

∎

At this moment, you might ask, What does this problem have to do with generating functions? How do we use generating functions? Our answer to the first question: What if we are asking for $(0,0)$ and $(50, 70)$, instead? In fact, we want to find a good way to compute

$$\sum_{i=0}^{n-1} \binom{n}{i} \cdot \binom{n}{i+1}.$$

Our answer to the second question is the following. Consider $f(x) = (1+x)^{2n}$. Recall that for the polynomial $p(x)$, $[x^m]P(x)$ denotes the coefficient of x^m in the expansion of $p(x)$. Thus, $[x^{n-1}]f(x) = \binom{2n}{n-1}$. On the other hand, we have

$$f(x)$$
$$= (1+x)^2 = (a_0 + a_1 x + a_2 x^2 + \cdots + a_n x^n)^2$$
$$= (a_0 + a_1 x + a_2 x^2 + \cdots + a_n x^n)(a_0 + a_1 x + a_2 x^2 + \cdots + a_n x^n)$$
$$= b_0 + b_1 x + b_2 x^2 + \cdots + b_{2n} x^{2n},$$

where $a_i = \binom{n}{i}$ for $0 \le i \le n$. By the rules of polynomial multiplication,

$$[x^{n-1}]f(x) = b_{n-1} = a_0 a_{n-1} + a_1 a_{n-2} + \cdots + a_{n-1} a_0$$

$$= \sum_{i=0}^{n-1} \binom{n}{i} \cdot \binom{n}{i+1}.$$

Thus, by calculating the coefficient of x^{n-1} in $f(x)$ in two ways, we obtain

$$\sum_{i=0}^{n-1} \binom{n}{i} \cdot \binom{n}{i+1} = \binom{2n}{n-1}. (*)$$

In light of the above discussion, by setting $n = 6$, we conclude that

$$\sum_{i=0}^{5} a_i b_i = \sum_{i=0}^{5} \binom{6}{i} \cdot \binom{6}{i+1} = \binom{12}{5}.$$

This result can also be obtained via a bijection. Consider the $\binom{12}{5}$ walks that start at $(0,0)$, end at $(5,7)$, and consist of 12 steps, each one either up or to the right. There is a one-to-one correspondence between these walks and the set of (A, B)-paths where A and B meet. In particular, given one of the $\binom{12}{5}$ walks from $(0,0)$ to $(5,7)$, the path followed by A consists of the first six steps of the walk, and the path followed by B is obtained by starting at $(5,7)$ and reversing the last six steps of the walk. It is not difficult to see that this model can be generalized. The readers might have already noticed that identity $(*)$ is a special case of the Vandermonde Identity (Example 3.10):

$$\binom{m+n}{k} = \sum_{i=0}^{k} \binom{m}{i}\binom{n}{k-i},$$

where m, n, and k are integers, with $m, n \geq 0$. The above identity can be proven by calculating the coefficient of x^k in the expansion of $(1+x)^{m+n}$ in two ways. We will leave this as an exercise.

The above example showed a typical technique in applying generating functions – obtaining identities by calculating the coefficients of certain terms of a polynomial in two ways. Another typical technique involves expanding functions into infinite polynomial series (or power series). Thus, we will clarify the operation rules of power series before we apply them.

If $A = (a_0, a_1, a_2, \dots)$ and $B = (b_0, b_1, b_2, \dots)$ are two infinite series, then

$$P_A(x) + P_B(x) = \sum_{n=0}^{\infty}(a_n + b_n)x^n = P_{A+B}(x),$$

where $A + B = (a_0 + b_0, a_1 + b_1, a_2 + b_2, \dots)$. Furthermore, for a constant α, we define the sequence

$$\alpha A = (\alpha a_0, \alpha a_1, \alpha a_2, \dots).$$

Let

$$C = (c_0, c_1, c_2, \dots, c_n, \dots),$$

where $c_0 = a_0 b_0$, $c_1 = a_0 b_1 + a_1 b_0$, $c_2 = a_0 b_2 + a_1 b_1 + a_2 b_0$, and in general, $c_n = \sum_{k=0}^{n} a_k b_{n-k}$. We have

$$\alpha P_A(x) = \alpha \sum_{n=0}^{\infty} a_n x^n = \sum_{n=0}^{\infty} \alpha a_n x^n = P_{\alpha A}(x)$$

and

$$P_A(x)P_B(x) = \left(\sum_{n=0}^{\infty} a_n x^n\right)\left(\sum_{n=0}^{\infty} b_n x^n\right)$$

$$= \sum_{n=0}^{\infty}\left(\sum_{k=0}^{n} a_k b_{n-k}\right) x^n$$

$$= P_C(x).$$

Note that the arithmetic operations of infinite power series and finite power series are defined in exactly the same way. For example, $\alpha(a_0 + a_1 x + a_2 x^2) = \alpha a_0 x + \alpha a_1 x + \alpha a_2 x^2$ and

$$(a_0 + a_1 + a_2 x^2)(b_0 + b_1 x)$$

$$= a_0 b_0 + (a_0 b_1 + a_1 b_0)x + (a_1 b_1 + a_2 b_0)x^2 + (a_2 b_1)x^3,$$

where we can express $a_2 b_1 = a_0 b_3 + a_1 b_2 + a_2 b_1 + a_3 b_0$ by setting $a_3 = b_2 = b_3 = 0$. Note also that by setting $\alpha = -1$, the series $P_{-A}(x) = -P_A(x)$ and $P_{A-B}(x) = P_A(x) - P_B(x) = P_A(x) + (-P_B(x))$ are well-defined.

Proposition 8.1. *Let A and B be two sequences with*

$$A = (a_0, a_1, a_2, \dots),$$

$$B = (b_0, b_1, b_2, \dots) = (a_0, a_0 + a_1, a_0 + a_1 + a_2, \dots),$$

$$C = (c_0, c_1, c_2, \dots) = (1, 1, 1, \dots),$$

$$D = (d_0, d_1, d_2, \dots),$$

where $d_n = \sum_{k=0}^{n} a_k a_{n-k}$. Then $P_A(x)P_C(x) = P_B(x)$ and $(P_A(x))^2 = P_D(x)$.

Proof: By the definition of the product of two power series, we have

$$P_A(x)P_C(x) = \sum_{n=0}^{\infty}\left(\sum_{k=0}^{n} a_k c_{n-k}\right) x^n = \sum_{n=0}^{\infty}\left(\sum_{k=0}^{n} a_k\right) x^n = P_B(x)$$

and

$$(P_A(x))^2 = P_A(x)P_A(x) = \sum_{n=0}^{\infty}\left(\sum_{k=0}^{n} a_k a_{n-k}\right) x^n = P_D(x),$$

as desired. ∎

Example 8.2.   If $A = (1,1,1,\dots)$ and $B = (1,2,3,\dots)$, find $P_A(x)$ and $P_B(x)$.

Solution: Let $C = (c_0, c_1, c_2, \dots) = (1, -1, 0, 0, \dots)$.   Then $P_C(x) = 1 - x$ and $P_A(x) = 1 + x + x^2 + \dots$. Let $A = (a_0, a_1, a_2, \dots)$. Then $a_0 c_0 = 1$, and

$$\sum_{k=0}^{n} a_k c_{n-k} = a_0 c_n + a_1 c_{n-1} + \dots + a_n c_0 = a_{n-1} c_1 + a_n c_0 = 0,$$

for $n \geq 1$. Hence $(1 - x)P_A(x) = 1$, or

$$P_A(x) = 1 + x + x^2 + \dots = \frac{1}{1-x}.$$

Note that

$$B = (a_0, a_0 + a_1, a_0 + a_1 + a_2, \dots).$$

We have

$$P_B(x) = 1 + 2x + 3x^2 + \dots = (1 + x + x^2 + \dots)^2 = \frac{1}{(1-x)^2}.$$

∎

The results above can easily be derived as **Maclaurin series** of functions in calculus. On the other hand, we are not considering the convergence of those series. It is important to note that because we do not consider the convergence of polynomials with infinitely many terms, it is not meaningful to compute the value of a generating function at certain values. For example, for $x \geq 1$,

$$1 + x + x^2 + \dots \neq \frac{1}{1-x}.$$

We will revisit Examples 3.8, 4.11, and 5.13 to illustrate the applications of Maclaurin series and generating functions.

Example 8.3.   Let $\{F_n\}_{n=0}^{\infty}$ be the Fibonacci sequence. Determine an explicit formula for F_n.

Solution: Let $F = (F_0, F_1, F_2, \dots)$, where $F_0 = F_1 = 1$ and $F_{n+2} = F_{n+1} + F_n$ for $n \geq 0$. Then

$$
\begin{aligned}
P_A(x) &= 1 &+x &+2x^2 &+3x^3 &+5x^4 + \cdots, \\
-xP_A(x) &= &-x &-x^2 &-2x^3 &-3x^4 - \cdots, \\
-x^2 P_A(x) &= & &-x^2 &-x^3 &-2x^4 - \cdots,
\end{aligned}
$$

implying that $P_A(x) - xP_A(x) - x^2 P_A(x) = 1$, or

$$P_A(x) = \frac{1}{1 - x - x^2}.$$

The quadratic equation $x^2 + x - 1 = 0$ has the roots $r_1 = \frac{-1+\sqrt{5}}{2}$ and $r_2 = \frac{-1-\sqrt{5}}{2}$. Thus, $(r_1 - x)(r_2 - x) = (x - r_1)(x - r_2) = x^2 + x_1 - 1 = -(1 - x - x^2)$. Therefore,

$$P_A(x) = \frac{1}{r_1 - r_2} \cdot \frac{r_1 - r_2}{1 - x - x^2} = \frac{1}{r_1 - r_2} \cdot \frac{r_1 - r_2}{-(r_1 - x)(r_2 - x)}$$

$$= \frac{1}{r_1 - r_2} \cdot \frac{(r_1 - x) - (r_2 - x)}{-(r_1 - x)(r_2 - x)} = \frac{1}{\sqrt{5}} \left(\frac{1}{r_1 - x} - \frac{1}{r_2 - x} \right),$$

by noting that $r_1 - r_2 = \sqrt{5}$. Setting $x' = \frac{x}{r_1}$ gives

$$\frac{1}{r_1 - x} = \frac{1}{r_1} \cdot \frac{1}{1 - x'} = \frac{1}{r_1} \sum_{n=0}^{\infty} (x')^n = \sum_{n=0}^{\infty} \frac{1}{r_1^{n+1}} x^n,$$

by Example 8.2. Similarly, we have

$$\frac{1}{r_2 - x} = \sum_{n=0}^{\infty} \frac{1}{r_2^{n+1}} x^n.$$

It follows that

$$\frac{1}{r_1 - x} - \frac{1}{r_2 - x} = \sum_{n=0}^{\infty} \left(\frac{1}{r_1^{n+1}} - \frac{1}{r_2^{n+1}} \right) x^n$$

$$= \sum_{n=0}^{\infty} \frac{1}{(r_1 r_2)^{n+1}} (r_2^{n+1} - r_1^{n+1}) x^n$$

$$= \sum_{n=0}^{\infty} (-1)^{n+1} (r_2^{n+1} - r_1^{n+1}) x^n,$$

because $r_1 r_2 = -1$. Hence,

$$P_A(x) = \frac{1}{\sqrt{5}} \sum_{n=0}^{\infty} (-1)^{n+1} (r_2^{n+1} - r_1^{n+1}) x^n$$

$$= \sum_{n=0}^{\infty} \frac{1}{\sqrt{5}} [(-r_2)^{n+1} - (-r_1)^{n+1}] x^n.$$

Therefore,

$$F_n = [x^n]P_A(x) = \frac{1}{\sqrt{5}}[(-r_2)^{n+1} - (-r_1)^{n+1}]$$

$$= \frac{1}{\sqrt{5}}\left[\left(\frac{1+\sqrt{5}}{2}\right)^{n+1} - \left(\frac{1-\sqrt{5}}{2}\right)^{n+1}\right].$$

∎

The above formula can certainly be derived by using Theorems 5.1 and 5.2. This approach reveals the connection between generating functions and characteristic equations of recursive relations.

Example 8.4.   Let n be a positive integer with $n \geq 3$. Each of n students has a different height. In how many ways can they line up in a row such that the heights of any three of them (not necessarily next to each other), from left to right, are not in the following order: medium, tall, short?

Solution: Let a_n denote the number of line-ups that satisfy the conditions of the problem, and let $A = (a_0, a_1, a_2, \ldots)$, with $a_0 = 1, a_1 = 1$, and $a_2 = 2$.

This is yet another combinatorial model of the Catalan numbers (Examples 4.11 and 5.13). We leave it to the reader as an exercise to show that

$$a_n = a_0 a_{n-1} + a_1 a_{n-2} + \cdots + a_{n-1} a_0.$$

Note that the right-hand side of the above identity is $\sum_{k=0}^{n} a_k a_{n-k}$, that is, the coefficient of x^{n-1} of $(P_A(x))^2$, by Proposition 8.1. In other words, for $n \geq 1$, the coefficient of x^{n-1} in $(P_A(x))^2$ is equal to the coefficient of the x^n term in $P_A(x)$. Thus, the series $x(P_A(x))^2$ matches the series $P_A(x)$ term by term except for the constant term. We have

$$x(P_A(x))^2 = P_A(x) - 1.$$

Solving the above equation as a quadratic in $P_A(x)$ gives

$$P_A(x) = \frac{1 \pm \sqrt{1 - 4x}}{2x}.$$

Next we expand $\sqrt{1-4x}$ as a power series. Let $f(x) = \sqrt{1-4x}$. Then it is not difficult to check that

$$f'(x) = -2(1-4x)^{-1/2},$$

$$f''(x) = -2^2(1-4x)^{-3/2},$$

$$f^{(3)}(x) = -2^3 \cdot 3(1-4x)^{-5/2},$$

$$f^{(4)}(x) = -2^4 \cdot 3 \cdot 5(1-4x)^{-7/2},$$

$$\cdots$$

$$f^{(n)}(x) = -2^n \cdot 3 \cdot 5 \cdots (2n-3)(1-4x)^{-(2n-1)/2}$$

$$= -2^n \cdot \frac{(2n-2)!}{2 \cdot 4 \cdots (2n-2)} \cdot (1-4x)^{-(2n-1)/2}$$

$$= -\frac{2(2n-2)!}{(n-1)!} \cdot (1-4x)^{-(2n-1)/2},$$

where for $n \geq 3$, $f^{(n)}(x)$ denotes the nth derivative of $f(x)$. Thus, the Maclaurin series for $\sqrt{1-4x}$ is

$$\sum_{n=0}^{\infty} \frac{f^n(0)}{n!} x^n = 1 - \sum_{n=1}^{\infty} \frac{2(2n-2)!}{(n-1)! \cdot n!} x^n = 1 - \sum_{n=1}^{\infty} \frac{2}{n} \binom{2n-2}{n-1} x^n.$$

Consequently,

$$1 \pm \sqrt{1-4x} = 1 \pm 1 \mp \sum_{n=1}^{\infty} \frac{2}{n} \binom{2n-2}{n-1} x^n.$$

Because the a_i are clearly nonnegative, we have

$$P_A(x) = \frac{1-\sqrt{1-4x}}{2x} = \frac{\sum_{n=1}^{\infty} \frac{2}{n}\binom{2n-2}{n-1}x^n}{2x}$$

$$= \sum_{n=1}^{\infty} \frac{1}{n} \binom{2n-2}{n-1} x^{n-1} = \sum_{n=0}^{\infty} \frac{1}{n+1} \binom{2n}{n} x^n,$$

implying that

$$a_n = \frac{1}{n+1} \binom{2n}{n},$$

as desired.   ∎

There are many more beautiful combinatorial identities that can be obtained with generating functions. On the other hand, most of those generating functions require more clarification on the division and radical operations of power series (and/or general definitions of binomial coefficients $\binom{m}{n}$, where m and n are complex numbers, using gamma functions), which are a bit too technical for this book. Interested readers are encouraged to read books or articles on these special topics.

Next, we will introduce another kind of generating function which deals with integer sequences.

If $A = (a_0, a_1, a_2, \cdots)$ is a integer sequence, then

$$E_A(x) = x^{a_0} + x^{a_1} + x^{a_2} + \cdots$$

is the **generating function** (of the second type) associated with A. If all of the entries of A are nonnegative, then $E_A(x)$ is a polynomial. Otherwise, $E_A(x)$ can be written as the quotient of two polynomials.

Generating functions of the second type effectively deal with the addition of two integer sequences.

Proposition 8.2. *Let $A = (a_1, a_2, \ldots, a_m)$ and $B = (b_1, b_2, \ldots, b_n)$ be two integer sequences. Then the sequence*

$$C = \{c_{i,j} \mid a_i + b_j, 1 \le i \le m, 1 \le j \le n\}$$

is associated with $E_C(x) = E_A(x)E_B(x)$. In particular, the coefficient of x^k in $E_C(x)$ is equal to the number of ordered pairs of integers (a_i, b_j) with $a_i \in A$ and $b_j \in B$ such that $a_i + b_j = k$.

Proof: Note that $E_A(x) = \sum_{i=1}^m x^{a_i}$ and $E_B(x) = \sum_{j=1}^n x^{b_j}$. Hence

$$E_A(x)E_B(x) = \sum_{i=1}^m x^{a_i} \cdot \sum_{j=1}^n x^{b_j} = \sum_{\substack{1 \le i \le m \\ 1 \le j \le n}} x^{a_i + b_j} = E_C(x),$$

where the last sum is taken over all (a, b) with $a \in A$ and $b \in B$. We also note that the term x^k appears in $E_C(x)$ if and only if there are $a_i \in A$ and $b_j \in B$ such that $k = a_i + b_j$, and that each of these pairs contributes exactly once to the number of appearances of x^k in $E_C(x)$. ∎

Consider the following problem. A bug makes some random moves along the x-axis. It starts at $(0, 0)$, and on each of its moves it may either move one unit right or stay put. In how many ways can it land on $(i, 0)$ after n moves? We may consider the sequence $A = (1, 0)$, where 1 and 0 represent the possible increments of the x-coordinate at each move. By applying Proposition 8.2 $n - 1$ times, $(E_A(x))^n =$

$(x^0+x^1)^n = (1+x)^n = \sum_{i=1}^{n} \binom{n}{i} x^i$ gives all of the desired information. We thus interpret $(1+x)^n$ – the most fundamental generating function of the first type – as a power of generating functions of the second type.

Example 8.5.   Claudia has two regular tetrahedra, and she decides to make them into two fair dice. On each face of the dice some positive integer is written such that the numbers 1, 2, 3, 4 do not all appear on one die. Claudia is also interested in the distribution of the sums of the numbers on the bottom faces when the dice are thrown. She wants this distribution to be the same as that of two tetrahedral dice when the numbers 1, 2, 3, 4 are written on the faces of each die. Can Claudia achieve her goal?

Solution: The answer is yes. Claudia can achieve her goal by using the numbers 1, 2, 2, 3 on one die and the numbers 1, 3, 3, 5 on the other.

Where are these answers from? Or how do we check if they work? To answer these questions, let's put generating functions of the second type into action.

Assume that Claudia uses the numbers a_1, a_2, a_3, a_4 on one die and the numbers b_1, b_2, b_3, b_4 on the other. Claudia needs to consider the possible sums of two numbers in the sequences $A = (a_1, a_2, a_3, a_4)$ and $B = (b_1, b_2, b_3, b_4)$ and their distribution. This distribution is identical to that of the possible sums of two numbers in the sequence $C = (1, 2, 3, 4)$. By Proposition 8.2, the conditions of the problem are satisfied if and only if

$$(x^{a_1} + x^{a_2} + x^{a_3} + x^{a_4})(x^{b_1} + x^{b_2} + x^{b_3} + x^{b_4}) = (x^1 + x^2 + x^3 + x^4)^2,$$

where $\{a_1, a_2, a_3, a_4\} \neq \{1, 2, 3, 4\}$. Note that

$$(x^1 + x^2 + x^3 + x^4)^2 = x^2(1+x)^2(1+x^2)^2$$
$$= x^2(1 + 2x + x^2)(1 + 2x^2 + x^4)$$
$$= (x + x^2 + x^2 + x^3)(x + x^3 + x^3 + x^5),$$

which answers both of our questions.   ∎

Now we visit a more complicated version of Example 4.6.

Example 8.6.   Six regular dice are rolled. What is the probability that the sum of the six numbers shown is equal to 21?

Solution: Let $d_1, d_2, \ldots, d_6$ denote the dice, and let x_i be the number shown on die d_i. We need to compute the number of ordered six-tuples of integers $(x_1, x_2, \ldots, x_6)$ with $1 \leq x_i \leq 6$ such that

$$x_1 + x_2 + x_3 + x_4 + x_5 + x_6 = 21.$$

By Proposition 8.2, this number is equal to the coefficient of x^{21} in the expansion of

$$(x + x^2 + x^3 + x^4 + x^5 + x^6)^6,$$

because $x + x^2 + x^3 + x^4 + x^5 + x^6$ is the generating function of the sequence $(1, 2, 3, 4, 5, 6)$. This coefficient can easily be obtained by the assistance of any computer algebra system. If we really want to, we can also compute this number by hand. This is a good exercise in applying Proposition 8.2, so we will leave it to the reader. ∎

The following example is another classical application of generating functions. For its relationship to the Morse sequence, please see [51].

Example 8.7.   Let $(a_1, a_2, \ldots, a_n)$ and $(b_1, b_2, \ldots, b_n)$ be two different unordered n-tuples of integers such that the sequences

$$a_1 + a_2, a_1 + a_3, \ldots, a_{n-1} + a_n \text{ (all pairwise sums } a_i + a_j, 1 \leq i < j \leq n)$$

and

$$b_1 + b_2, b_1 + b_3, \ldots, b_{n-1} + b_n \text{ (all pairwise sums } b_i + b_j, 1 \leq i < j \leq n)$$

coincide up to permutation. Prove that n is a power of two.

Solution: For constant c, the n-tuples $(c+a_1, c+a_2, \ldots, c+a_n)$ and $(c + b_1, c + b_2, \ldots, c + b_n)$ still satisfy the conditions of the problem. Thus, without loss of generality, we may assume that $(a_1, a_2, \ldots, a_n)$ and $(b_1, b_2, \ldots, b_n)$ are two n-tuples of positive integers. (This modification guarantees that all of the generating functions involved will be polynomials.)

Let $A = (a_1, a_2, \ldots, a_n)$ and $B = (b_1, b_2, \ldots, b_n)$. Then

$$E_A(x) = \sum_{i=1}^{n} x^{a_i} \quad \text{and} \quad E_B(x) = \sum_{i=1}^{n} x^{b_i}.$$

By Proposition 8.2, we have

$$(E_A(x))^2 = \sum_{k=1}^{n} x^{2k} + 2 \sum_{1 \le i < j \le n} x^{a_i + a_j} = E_A(x^2) + 2 \sum_{1 \le i < j \le n} x^{a_i + a_j}$$

and

$$(E_B(x))^2 = E_B(x^2) + 2 \sum_{1 \le i < j \le n} x^{b_i + b_j}.$$

Because the sequences

$$a_1 + a_2, a_1 + a_3, \ldots, a_{n-1} + a_n$$

and

$$b_1 + b_2, b_1 + b_3, \ldots, b_{n-1} + b_n$$

coincide up to permutation, we have

$$\sum_{1 \le i < j \le n} x^{a_i + a_j} = \sum_{1 \le i < j \le n} x^{b_i + b_j}.$$

Consequently,

$$(E_A(x))^2 - (E_B(x))^2 = E_A(x^2) - E_B(x^2).$$

Because the unordered n-tuples A and B are distinct, $E_A(x) - E_B(x) \ne 0$. Hence, we have

$$E_A(x) + E_B(x) = \frac{E_A(x^2) - E_B(x^2)}{E_A(x) - E_B(x)}.$$

Note that $E_A(1) = E_B(1) = n$, that is, $x = 1$ is a root of $E_A(x) - E_B(x)$ because both $E_A(x)$ and $E_B(x)$ are polynomials. Thus

$$E_A(x) - E_B(x) = (x - 1)^k f(x),$$

where k is a positive integer and $f(x)$ is a polynomial not divisible by $x - 1$, that is, $f(1) \ne 0$. Then $E_A(x^2) - E_B(x^2) = (x^2 - 1)^k f(x^2)$. It follows that

$$E_A(x) + E_B(x) = \frac{(x^2 - 1)^k f(x^2)}{(x - 1)^k f(x)} = \frac{(x + 1)^k f(x^2)}{f(x)}.$$

Setting $x = 1$ in the above identity gives

$$2n = E_A(1) + E_B(1) = \frac{2^k f(1)}{f(1)} = 2^k,$$

implying that $n = 2^{k-1}$, as desired.   ∎

Note that for a scalar $c \neq 0$ and n-tuples $(a_1, a_2, \ldots, a_n)$ and $(b_1, b_2, \ldots, b_n)$ satisfying the conditions of Example 8.7, the n-tuples $(ca_1, ca_2, \ldots, ca_n)$ and $(cb_1, cb_2, \ldots, cb_n)$ still satisfy the same conditions. Thus the above result can be generalized to two rational n-tuples.

The next example deals with weighted generating functions of the second type.

Example 8.8. [Putnam 2001] Adrian has the coins $C_1, C_2, \ldots, C_n$. For each k, C_k is biased so that, when tossed, it has probability $1/(2k + 1)$ of showing heads. If the n coins are tossed, what is the probability that the number of heads is odd? Express the answer as a rational function of n.

Solution: Since we only want to consider probabilities of obtaining heads, we set 1 for a head and 0 for a tail. For a regular coin, $x^0 + x^1$ is associated with the sequence $(0, 1)$ (tail, head). We can then use $(1+x)^n$ to investigate the distribution of heads and tails when n coins are tossed. Each coin C_k is associated with the sequence

$$(\underbrace{0, 0, \ldots, 0}_{2k}, 1),$$

because it is $2k$ times more likely to show tails than heads. Considering probabilities, the polynomial

$$\frac{1}{2k+1}x + \frac{2k}{2k+1} = \frac{1}{2k+1}(x + 2k)$$

is associated with C_k. Put

$$f(x) = \frac{1}{2+1}(x + 2) \cdot \frac{1}{4+1}(x + 4) \cdots \frac{1}{2n+1}(x + 2n) = \prod_{k=1}^{n} \frac{x + 2k}{2k + 1}.$$

Then the coefficient of x^i in $f(x)$ is the probability of getting exactly i heads. The desired probability is the sum of the coefficients of the odd powers of x in the expansion of $f(x)$. Note that $f(1)$ is the sum of all the coefficients, and that $f(-1)$ is equal to the sum of the coefficients of the even powers of x minus the sum of the coefficients of the odd powers of x. Thus the desired number is $p = \frac{f(1) - f(-1)}{2}$. Both $f(1)$ and $f(-1)$ can be computed directly: $f(1) = 1$, and

$$f(-1) = \frac{1}{3} \cdot \frac{3}{5} \cdots \frac{2n-1}{2n+1} = \frac{1}{2n+1}.$$

The desired probability is then

$$p = \frac{1}{2}\left(1 - \frac{1}{2n+1}\right) = \frac{n}{2n+1}.$$

∎

The above solution belongs to Richard Stanley. Substituting values into generating functions is not a practical idea in certain situations because of the convergence of the power series. But it was a perfect choice in the previous two examples. The reader will see more applications of this technique later in this chapter. Another solution of the problem applies a recursion. The Interested readers might want to find this recursion as an exercise.

Example 8.9.   [IMO 1998 Short–listed] Let $a_0, a_1, a_2, \ldots$ be an increasing sequence of nonnegative integers such that every nonnegative integer can be expressed uniquely in the form $a_i + 2a_j + 4a_k$, where i, j, and k are not necessarily distinct. Determine a_{1998}.

Solution: Note that $b_i + 2b_j + 4b_k = \overline{b_k b_j b_i}_{(2)}$, where $n_{(a)}$ denotes the number n in base a. Note also that $\{\overline{b_k b_j b_i}_{(2)}\} = \{0, 1, 2, \ldots, 7\}$. It make sense to think about base 8 representations of nonnegative integers, because all of the digits are $0, 1, 2, \ldots, 7$. Let $A = (a_0, a_1, a_2, \ldots)$ be an increasing sequence of nonnegative integers whose representations in base 8 contain only the digits 0 and 1. Then for an arbitrary nonnegative integer $n_{(8)} = \overline{n_m n_{m-1} \ldots n_1 n_0}$, each of its digits n_k, $0 \le k \le m$, can be written uniquely in the form $n_k = \overline{b_{k,2} b_{k,1} b_{k,0}}_{(2)}$ (because $0 \le n_k \le 7$), that is, $n_k = 4b_{k,2} + 2b_{k,1} + b_{k,0}$, where $b_{k,i} = 0$ or 1 for $i = 0, 1, 2$. Then $n_{(8)} = 4x_{(8)} + 2y_{(8)} + z_{(8)}$, where

$$x_{(8)} = \overline{b_{m,2} b_{m-1,2} \ldots b_{1,2} b_{0,2}},$$

$$y_{(8)} = \overline{b_{m,1} b_{m-1,1} \ldots b_{1,1} b_{0,1}},$$

$$z_{(8)} = \overline{b_{m,0} b_{m-1,0} \ldots b_{1,0} b_{0,0}}.$$

By the definition of A, it is clear that $x_{(8)}, y_{(8)}$, and $z_{(8)}$ are elements of A. Because the representation of n_k in the binary system is unique, we have shown that every nonnegative integer can be expressed uniquely in the form $a_i + 2a_j + 4a_k$, where a_i, a_j, and a_k and elements of A with i, j, and k not necessarily distinct. To determine a_{1998}, we first express 1998 in the binary system and then change the base to 8. Because $1998 = 11111001110_{(2)}$, we obtain $a_{1998} = 11111001110_{(8)}$.

It seems that the problem is solved, but is it? Not so fast, my friend! We have only found a sequence that works. But the problem certainly indicates that this sequence is unique, because otherwise, how do we know that there are no other possible values of a_{1998}? Thus, we will find a role for generating functions in our solution.

Assume that $A = (a_0, a_1, a_2, \ldots)$ is a sequence that satisfies the conditions of the problem. Then

$$E_A(x) = x^{a_0} + x^{a_1} + x^{a_2} + \cdots,$$

and consequently,

$$E_A(x^2) = x^{2a_0} + x^{2a_1} + x^{2a_2} + \cdots,$$

$$E_A(x^4) = x^{4a_0} + x^{4a_1} + x^{4a_2} + \cdots.$$

By Proposition 8.2, we have

$$E_A(x)E_A(x^2)E_A(x^4) = \sum_{i,j,k} x^{a_i + 2a_j + 4a_k}.$$

Because every nonnegative integer can be expressed uniquely in the form $a_i + 2a_j + 4a_k$, where a_i, a_j, a_k are elements of A with $i, j,$ and k not necessarily distinct, we have

$$\sum_{i,j,k} x^{a_i + 2a_j + 4a_k} = x^0 + x^{1'} + x^2 + \cdots = \frac{1}{1-x}.$$

Hence,

$$g(x) = F(x)F(x^2)F(x^4) = \frac{1}{1-x},$$

and consequently,

$$g(x^2) = F(x^2)F(x^4)F(x^8) = \frac{1}{1-x^2} = \frac{1}{(1-x)(1+x)}.$$

Thus,

$$\frac{F(x)}{F(x^8)} = 1 + x,$$

or $F(x) = (1 + x)F(x^8)$. Therefore,

$$\sum_{k=0}^{\infty} x^{a_k} = F(x) = (1+x)F(x^8) = (1+x)(1+x^8)F(x^{8^2})$$

$$= (1 + x)\left(1 + x^8\right)\left(1 + x^{8^2}\right)\left(1 + x^{8^3}\right)\cdots.$$

Hence the sequence A consists of all positive integers that are powers of 8 or sums of distinct powers of 8; that is, the a_is are integers in base 8 whose digits is 0 or 1. ∎

The reader might feel that the second part of the solution above (using generating functions) is already a complete solution to the problem, which is true. But nevertheless, we feel that the first part reveals the motivation for the problem.

The next three examples deal with partitions of positive integers.

Example 8.10.   [Putnam 1957] Let $\alpha(n)$ be the number of representations of a positive integer n as sum of 1's and 2's, taking order into account. For example, since

$$4 = 1+1+2 = 1+2+1 = 2+1+1 = 2+2 = 1+1+1+1,$$

we have $\alpha(4) = 5$. Let $\beta(n)$ be the number of representations of n that are sums of integers greater than 1, again taking order into account. For example, since

$$6 = 4+2 = 2+4 = 3+3 = 2+2+2,$$

we have $\beta(6) = 5$. Show that $\alpha(n) = \beta(n+2)$.

Solution: For each positive integer k, let $\alpha(n,k)$ denote the number of ordered k-tuples of integers $(a_1, a_2, \ldots, a_k)$ with $1 \le a_i \le 2$ such that $a_1 + a_2 + \cdots + a_k = n$. Let $A = (1,2)$. By repeatedly applying Proposition 8.2, we have

$$\alpha(n,k) = [x^n](x+x^2)^k.$$

It is clear that $\alpha(n) = \alpha(n,1) + \alpha(n,2) + \cdots$. (Of course, this is a finite sum, because $\alpha(n,k) = 0$ for large k.) Because $n \ge 1$, we know that $k \ge 1$. We also may assume that $\alpha(n,0) = 0$, which is equal to the coefficient of x^n in $(x+x^2)^0 = 1$. Hence for $n \ge 1$, $\alpha(n)$ is equal to the coefficient of x^n in $f(x)$, where

$$f(x) = 1 + (x+x^2) + (x+x^2)^2 + \cdots = \frac{1}{1-(x+x^2)} = \frac{1}{1-x-x^2}$$

by Proposition 8.1, or

$$1 + \sum_{n=1}^{\infty} \alpha(n)x^n = \frac{1}{1-x-x^2}.$$

Similarly, for $n \geq 2$ we can show that $\beta(n)$ is equal to the coefficient of x^n in $g(x)$, where

$$
\begin{aligned}
g(x) &= 1 + (x^2 + x^3 + \cdots) + (x^2 + x^3 + \cdots)^2 + \cdots \\
&= \frac{1}{1 - (x^2 + x^3 + \cdots)} = \frac{1}{1 - x^2(1 + x + \cdots)} \\
&= \frac{1}{1 - \frac{x^2}{1-x}} = \frac{1-x}{1-x-x^2} \\
&= 1 + \frac{x^2}{1 - x - x^2},
\end{aligned}
$$

by Proposition 8.1. Thus,

$$
1 + \sum_{n=2}^{\infty} \beta(n)x^n = 1 + \frac{x^2}{1 - x - x^2}.
$$

Therefore,

$$
\sum_{n=2}^{\infty} \beta(n)x^n = x^2 \left(1 + \sum_{n=1}^{\infty} \alpha(n)x^n \right) = x^2 + \sum_{n=1}^{\infty} \alpha(n)x^{n+2},
$$

implying that $\alpha(n) = \beta(n+2)$, as desired. ∎

This problem can also be solved using recursion or a bijection. The interested readers might want to find these solutions.

Example 8.11. [China 1999] For a set A, let $s(A)$ denote the sum of the elements of A. (If $A = \emptyset$, then $|A| = s(A) = 0$.) Let

$$
S = \{1, 2, \ldots, 1999\}.
$$

For $r = 0, 1, 2, \ldots, 6$, define

$$
T_r = \{T \mid T \subseteq S, \; s(T) \equiv r \pmod 7\}.
$$

For each r, find the number of elements in T_r.

Solution: If an integer i, $1 \leq i \leq 1999$, is in T, it contributes i in the sum $s(T)$; otherwise, it contributes 0. Hence for each number i, we associate it with the generating function $x^0 + x^i = 1 + x^i$. Consider the polynomial

$$
f(x) = (1 + x)(1 + x^2) \cdots (1 + x^{1999}) = \sum_n c_n x^n.
$$

Then there is a bijection between each subset $T = \{a_1, a_2, \ldots, a_m\}$ of S and each term $x^{a_1} x^{a_2} \cdots x^{a_m} = x^{a_1 + a_2 + \cdots + a_m}$. Hence

$$|T_r| = \sum_k [x^{7k+r}] f(x) = \sum_k c_{7k+r}.$$

Let $\xi = e^{\frac{2\pi}{7} i}$, where $i^2 = -1$, be a 7th root of unity. Then $\xi^7 = 1$ and $\xi \neq 1$, so ξ is a root of

$$\frac{x^7 - 1}{x - 1} = 1 + x + x^2 + \cdots x^6,$$

That is, $1 + \xi + \xi^2 + \cdots + \xi^6 = 0$. For r divisible by 7, we have $\sum_{k=1}^{6} \xi^{kr} = \sum_{k=1}^{6} 1 = 6$. For r not divisible by 7,

$$\{1, 2, \ldots, 6\} \equiv \{r \cdot 1, r \cdot 2, \cdots, r \cdot 6\} \pmod{7}.$$

(In other words, a complete set of residue classes modulo 7 remains invariant by multiplying r by each number in the set.) Thus,

$$\sum_{k=1}^{6} \xi^{kr} = \begin{cases} 6, & r \text{ is divisible by 7,} \\ -1, & r \text{ is not divisible by 7.} \end{cases} \quad (**)$$

Hence,

$$\sum_{i=0}^{6} f(\xi^i) = \sum_{i=0}^{6} \sum_n c_n \xi^{ni} = \sum_n c_n \sum_{i=0}^{6} \xi^{ni} = \sum_{7|n} 7 c_n = 7|T_0|.$$

In exactly the same way, we can show that

$$|T_r| = \frac{1}{7} \sum_{i=0}^{6} \xi^{-ri} f(\xi^i) = \frac{1}{7} \left(2^{1999} + \sum_{i=1}^{6} \xi^{-ri} f(\xi^i) \right),$$

since $f(\xi^0) = f(1) = 2^{1999}$. Note also that $\xi, \xi^2, \ldots, \xi^7 = 1$ are the roots of $g(x) = x^7 - 1$, that is,

$$g(x) = x^7 - 1 = (x - \xi)(x - \xi^2) \cdots (x - \xi^7).$$

It follows that

$$g(-1) = -2 = (-1 - \xi)(-1 - \xi^2)(-1 - \xi^3) \cdots (-1 - \xi^7),$$

implying that

$$(1 + \xi)(1 + \xi^2) \cdots (1 + \xi^7) = 2.$$

Consequently, because $1999 = 7 \cdot 285 + 4$, we have

$$
\begin{aligned}
f(\xi) &= (1 + \xi)(1 + \xi^2) \cdots (1 + \xi^{1999}) \\
&= [(1 + \xi)(1 + \xi^2) \cdots (1 + \xi^7)]^{285}(1 + \xi)(1 + \xi^2)(1 + \xi^3)(1 + \xi^4) \\
&= 2^{285} \cdot [(1 + \xi)(1 + \xi^2)(1 + \xi^4)](1 + \xi^3) \\
&= 2^{285} \cdot (1 + \xi + \xi^2 + \cdots + \xi^7)(1 + \xi^3) \\
&= 2^{285}(1 + \xi^3).
\end{aligned}
$$

In general, we have $f(\xi^i) = 2^{285}(1 + \xi^{3i})$ for $1 \le i \le 6$. It follows that

$$
\begin{aligned}
|T_r| &= \frac{1}{7}\left(2^{1999} + 2^{285} \sum_{i=1}^{6} \xi^{-ri}(1 + \xi^{3i}) \right) \\
&= \frac{1}{7}\left(2^{1999} + 2^{285} \sum_{i=1}^{6} \left[\xi^{-ri} + \xi^{(3-r)i} \right] \right)
\end{aligned}
$$

By equation $(**)$, we conclude that

$$
\sum_{i=1}^{6} \left[\xi^{-ri} + \xi^{(3-r)i} \right] = \begin{cases} 6 - 1 = 5, & r \equiv 0 \text{ or } 3 \pmod 7, \\ -1 - 1 = -2, & \text{otherwise.} \end{cases}
$$

Therefore, the answer to the problem is

$$
|T_r| = \begin{cases} \dfrac{2^{1999} + 5 \cdot 2^{285}}{7} & r = 0 \text{ or } 3, \\[2mm] \dfrac{2^{1999} - 2^{286}}{7} & r = 1, 2, 4, 5, 6. \end{cases}
$$

∎

It is not hard to see that the above method can be generalized to $S = \{1, 2, \ldots, p\}$ for any prime p. A somewhat related but more difficult problem appeared on the IMO in 1995, and can be solved using generating functions in two variables.

Example 8.12.   [IMO 1995] Let p be an odd prime number. How many p-element subsets A of $\{1, 2, \ldots, 2p\}$ are there such that the sums of its elements are divisible by p?

Solution: For a number i, $1 \le i \le 2p$, we cannot simply associate it with $x^0 + x^i = 1 + x^i$, because the product

$$
\prod_{i=1}^{2p}(1 + x^i)
$$

cannot control the condition that the subsets have p elements. (For example, the coefficient of x^{kp} is equal to the number of subsets with the sum of its elements equal to kp, but many of these subsets do not have exactly p elements.)

Instead, we consider the generating function

$$g(t,x) = (t+x)(t+x^2)(t+x^3)\cdots(t+x^{2p}) = \sum_{k,m} a_{k,m} l^k x^m.$$

Then $a_{k,m}$ is equal to the number of subsets S of $\{1,2,\ldots,2p\}$ such that

(i) $|S| = 2p - k$;

(ii) the sum of the elements of S is m.

Thus, the answer to the problem is

$$A = \sum_{p|m} a_{p,m}.$$

Note that

$$\sum_{p|k,p|m} a_{k,m} = \sum_{p|m} a_{p,m} + \sum_{p|m} a_{0,m} + \sum_{p|m} a_{2p,m} = \sum_{p|m} a_{p,m} + 2,$$

because there is only one subset with each of $2p$ or 0 elements. It is easier to compute

$$B = \sum_{p|k,p|m} a_{k,m} = A + 2.$$

In order to find B, we use roots of unity as we did in Example 8.11. Let $\xi = e^{\frac{2\pi}{p}i}$, where $i^2 = -1$, be a pth root of unity. Let

$$E = \{\xi, \xi^2, \ldots, \xi^{p-1}, \xi^p = 1\}.$$

We compute

$$\sum_{t\in E}\sum_{x\in E} g(t,x)$$

in two ways. (Note that here we compute a double sum over all p^{th} roots of unity instead of a single sum of those roots.)

First, we have

$$\sum_{x\in E} g(t,x) = g(t,1) + \sum_{x\in E/\{1\}} g(t,x) = g(t,1) + \sum_{1\leq i\leq p-1} g(t,\xi^i).$$

It is clear that $g(t, 1) = (t+1)^{2p}$. As in the solution to Example 8.11, we have

$$\{1, 2, \cdots, p\} \equiv \{r \cdot 1, r \cdot 2, \cdots, r \cdot p\} \pmod{p} \qquad (\dagger)$$

for integers r not divisible by p. Consequently, for ξ^i with $1 \leq i \leq p - 1$,

$$(t + \xi^i)(t + \xi^{2i}) \cdots (t + \xi^{pi}) = (t + \xi)(t + \xi^2) \cdots (t + \xi^p),$$

implying that

$$g(t, \xi^i) = [(t + \xi)(t + \xi^2) \cdots (t + \xi^p)]^2 \quad \text{for } 1 \leq i \leq p - 1.$$

Also as in the solution to Example 8.11, we have

$$h(t) = (t - \xi)(t - \xi^2) \cdots (t - \xi^p) = t^p - 1,$$

implying that

$$h(-t) = (-t - \xi)(-t - \xi^2) \cdots (-t - \xi^p) = (-t)^p - 1 = -(t^p + 1).$$

It follows that

$$g(t, \xi^i) = [(t + \xi)(t + \xi^2) \cdots (t + \xi^p)]^2 = [(-1)^p h(-t)]^2 = (t^p + 1)^2.$$

Combining the above gives

$$\sum_{x \in E} g(t, x) = (t + 1)^{2p} + (p - 1)(t^p + 1)^2.$$

Thus, we have

$$\sum_{t \in E} \sum_{x \in E} g(t, x) = \sum_{t \in E} \left[(t+1)^{2p} + (p-1)(t^p+1)^2 \right]$$

$$= \sum_{t \in E} (t+1)^{2p} + (p-1) \sum_{t \in E} (t^p+1)^2$$

$$= \sum_{t \in E} \sum_{i=0}^{2p} \binom{2p}{i} t^i + 4p(p-1)$$

$$= \sum_{t \in E} \sum_{i=0,p,2p} \binom{2p}{i} t^i + \sum_{t \in E} \sum_{\substack{i \neq p \\ 0 < i < 2p}} \binom{2p}{i} t^i + 4p(p-1)$$

$$= p \left[2 + \binom{2p}{p} \right] + 4p(p-1)$$

$$= p \left[\binom{2p}{p} + 4p - 2 \right],$$

because

$$\sum_{t \in E} \sum_{\substack{i \neq p \\ 0 < i < 2p}} \binom{2p}{i} t^i$$

$$= \sum_{t \in E} \sum_{1 \le i \le p-1} \binom{2p}{i} t^i + \sum_{t \in E} \sum_{p+1 \le i \le 2p-1} \binom{2p}{i} t^i$$

$$= \sum_{1 \le i \le p-1} \binom{2p}{i} \sum_{t \in E} t^i + \sum_{p+1 \le i \le 2p-1} \binom{2p}{i} \sum_{t \in E} t^i = 0.$$

Note that in the above calculations, we used the fact that

$$\sum_{x \in E} x^r = \sum_{i=0}^{p-1} \xi^{ri} = \begin{cases} 0, & \text{if } r \text{ is not divisible by } p \\ p, & \text{if } r \text{ is divisible by } p, \end{cases} \tag{‡}$$

which follows from

$$1 + \xi + \xi^2 + \cdots + \xi^{p-1} = 0$$

and relation (†).

On the other hand, because $g(t, x) = \sum_{k,m} a_{k,m} t^k x^m$, we have

$$\sum_{t \in E} \sum_{x \in E} g(t, x)$$

$$= \sum_{t \in E} \sum_{x \in E} \sum_{k,m} a_{k,m} t^k x^m = \sum_{t \in E} \left[\sum_{k,m} a_{k,m} t^k \sum_{x \in E} x^m \right]$$

$$= \sum_{t \in E} \left[\sum_{k,m,p|m} a_{k,m} t^k \right] p = p \sum_{t \in E} \left[\sum_{k,m,p|m} a_{k,m} t^k \right]$$

$$= p \sum_{k,m,p|m} a_{k,m} \left[\sum_{t \in E} t^k \right] = p \sum_{k,m,p|k,p|m} a_{k,m} p$$

$$= p^2 B,$$

by applying relation (‡) repeatedly. Therefore, we obtain

$$p^2 B = \sum_{t \in E} \sum_{x \in E} g(t, x) = p \left[\binom{2p}{p} + 4p - 2 \right],$$

or

$$p(A + 2) = \binom{2p}{p} + 4p - 2.$$

Thus the answer to the problem is

$$A = \frac{1}{p} \left[\binom{2p}{p} - 2 \right] + 2.$$

■

We will discuss two more examples using generating functions in two variables.

Example 8.13. Let n be a positive integer, and let

$$f(x) = \sum_{k=0}^{n} \binom{n}{k}^2 (1 + x)^{2n-2k} (1 - x)^{2k}.$$

Show that $[x^{2m-1}] f(x) = 0$ for all positive integers m.

It seems that this problem can be solved directly by noting that

$$(1 + x)^{2n-2k} = \sum_{i=0}^{2n-2k} \binom{2n - 2k}{i} x^i$$

and

$$(1-x)^{2k} = \sum_{i=0}^{2k} (-1)^i \binom{2k}{i} x^i.$$

By Proposition 8.1, we have

$$(1+x)^{2n-2k}(1-x)^{2k} = \sum_{\ell=0}^{2n} \left(\sum_{i=0}^{\ell} (-1)^i \binom{2k}{i} \binom{2n-2k}{\ell-i} \right) x^\ell.$$

Thus,

$$f(x) = \sum_{k=0}^{n} \left[\binom{n}{k}^2 \sum_{\ell=0}^{2n} \left(\sum_{i=0}^{\ell} (-1)^i \binom{2k}{i} \binom{2n-2k}{\ell-i} \right) x^\ell \right]$$

$$= \sum_{k=0}^{n} \sum_{\ell=0}^{2n} \left[\sum_{i=0}^{\ell} (-1)^i \binom{n}{k}^2 \binom{2k}{i} \binom{2n-2k}{\ell-i} \right] x^\ell$$

$$= \sum_{\ell=0}^{2n} x^\ell \left[\sum_{k=0}^{n} \sum_{i=0}^{\ell} (-1)^i \binom{n}{k}^2 \binom{2k}{i} \binom{2n-2k}{\ell-i} \right].$$

It suffices to show that

$$\sum_{k=0}^{n} \sum_{i=0}^{\ell} (-1)^i \binom{n}{k}^2 \binom{2k}{i} \binom{2n-2k}{\ell-i} = 0 \quad \text{for } \ell \text{ odd}. \qquad (\dagger)$$

The only trouble is that identity ($\dagger$) is not easy to establish. Instead, we need to apply Proposition 8.1 in the opposite way!

Solution: Note that

$$\sum_{k=0}^{n} \binom{n}{k}^2 (1+x)^{2n-2k}(1-x)^{2k}$$

$$= \sum_{k=0}^{n} \binom{n}{k} [(1-x)^2]^k \binom{n}{n-k} [(1+x)^2]^{n-k}$$

is itself the coefficient of a certain term of in product of two power series. In particular, it is the coefficient of y^n in the expansion of

$$g(x,y) = [y + (1+x)^2]^n [y + (1-x)^2]^n.$$

Note that

$$g(x,y) = [y + (1+x)^2]^n [y + (1-x)^2]^n$$
$$= [(y + 1 + x^2 + 2x)(y + 1 + x^2 - 2x)]^n$$
$$= [(y + 1 + x^2)^2 - 4x^2]^n,$$

which is an even function in x. Thus, there are no odd-powered terms in x in the expansion of $g(x, y)$. Consequently, the coefficient of y^n in the expansion of $g(x, y)$ has only even-powered terms in x, proving the desired result. ∎

Thus we have also proved the interesting combinatorial identity (†).

Example 8.14. [A.M.M., 2002, by Emeric Deutsch] The bottom right unit square is removed from a 2×2 unit grid to form a *tromino*. Find the number of ways that k trominoes can be placed, without being rotated and without overlapping, on a $3 \times n$ rectangular unit grid. (Each tromino covers exactly three unit squares on the grid.)

Solution: After we have finished placing all k trominoes, we continue to tile the rectangle with unit squares. We call this process a *k-tromino tiling*. Let $a_{n,k}$ be the number of distinct k-tromino tilings of a $3 \times n$ rectangle. Instead of tiling the $3 \times n$ grid using trominoes and unit squares, we tile it with the four basic tiles shown in Figure 8.1

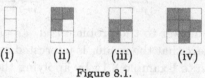

(i)   (ii)   (iii)   (iv)

Figure 8.1.

It is not difficult to see that a tiling using trominoes and unit squares is a concatenation of basic tiles and vice versa, and that such a sequence of basic tiles is unique in order from left to right. In other words, there is a bijection between the ways of tiling using trominoes and that of tiling using the four basic tiles.

We associate the term c with the basic tile (i), tc^2 with either (ii) or (iii), and t^2c^3 with (iv). (The reader might have already noticed the reason for choosing t and c: t stands for the number of trominoes, and c for the number of columns.) We do not consider the length of the rectangular grid for the moment; that is, we use the basic tiles to tile a rectangle with width 3 and an arbitrary length. We can associate one basic tile with the polynomial $p(c,t) = c + 2tc^2 + t^2 c^3 = c(1+tc)^2$, and j tiles with $p(c,t)^j = c^j (1+tc)^{2j}$, where j is a nonnegative integer. Then $p(c,t)^j$ describes all possible ways of tiling a rectangle with width 3 using j basic tiles.

Assume that a k-tromino tiling of a $3 \times n$ rectangle is a concatenation of j basic tiles: k_1 of type (i), k_2 of type (ii), k_3 of type (iii), and k_4 of type (iv). Then we have

$$1 \cdot k_1 + 2 \cdot k_2 + 2 \cdot k_3 + 3 \cdot k_4 = n,$$

$$0 \cdot k_1 + 1 \cdot k_2 + 1 \cdot k_3 + 2 \cdot k_4 = k.$$

Subtracting the second equation from the first yields $k_1 + k_2 + k_3 + k_4 = n - k$; that is, a k-tromino tiling of a $3 \times n$ rectangle is a concatenation of $j = n - k$ basic tiles. Thus we need to consider only the coefficients of the generating function $p(c,t)^{n-k} = c^{n-k}(1+tc)^{2(n-k)}$. In particular, we have

$$a_{n,k} = [t^k c^n] p(c,t)^{n-k} = [t^k c^k](1+tc)^{2(n-k)} = \binom{2n-2k}{k}.$$

∎

A shorter version of the above discussion is the following. Let $G(c,t) = \sum_{n,k \geq 0} a_{n,k} c^n t^k$. Then

$$G(c,t) = \sum_{j=0}^{\infty} p(c,t)^j = \sum_{j=0}^{\infty} c^j (1+tc)^{2j} = \sum_{j=0}^{\infty} \sum_{i=0}^{2j} \binom{2j}{i} c^{i+j} t^i,$$

implying that the answer to the problem is $a_{n,k} = \binom{2n-2k}{k}$.

By now, we believe that the reader is interested and able enough to find a nice solution to Example 8.14 by applying the bijection shown in Figure 8.2:

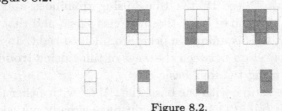

to

Figure 8.2.

Exercises 8

8.1. A bug makes some random moves along the x-axis.

  (a) It starts at $(0,0)$, and each of its moves is either a step right of length one or a step left of length one. In how many ways can it land on $(i,0)$ after n moves?

  (b) It starts at $(0,0)$, and each of its moves is either a step right of length one, a step left of length one, or staying put. In how many ways can it land on $(i,0)$ after n moves?

  (c) It starts at $(0,0)$, and each of its moves is either a step right of length one, a step left of length one, or staying put. The bug has just finished his lunch, so he is more likely to stay put on each move. In particular, it is equally likely for him to go either right or left at each move, but it is twice as likely that he stays put than that that he chooses to go right (or left) on each move. Compute the probability that the bug lands on $(i,0)$ after n moves.

8.2. Revisit the Vandermonde Identity (Example 3.10). Let m,n, and k be integers with $m,n \geq 0$. Prove that

$$\binom{m+n}{k} = \sum_{i=0}^{k} \binom{m}{i}\binom{n}{k-i}$$

by computing $[x^k](1+x)^{m+n}$ in two ways.

8.3. Compute the coefficient of x^{21} in the expansion of

$$(x + x^2 + x^3 + x^4 + x^5 + x^6)^6.$$

8.4. Adrian tosses 2003 fair coins, Andrea and Claudia each toss 2004 fair coins, and Zachary tosses 2005 fair coins. Show that the two events are equally likely, and determine the probability.

  (i) Claudia gets exactly one more head than Andrea does;

  (ii) Adrian and Zachary get exactly the same number of heads

8.5. Express

$$\binom{n}{0}^2 - \binom{n}{1}^2 + \binom{n}{2}^2 - \cdots + (-1)^n\binom{n}{n}^2$$

in closed form.

8.6. Let p be a prime. Compute the number of subsets T of $\{1, 2, \ldots, p\}$ such that p divides the sum of the elements of T.

8.7. Two more problems on dice:

(a) On each face of two dice some positive integer is written. The two dice are thrown, and the numbers on the top faces are added. Determine whether one can select the integers on the faces so that the possible sums are $2, 3, \ldots, 13$, all equally likely.

(b) A die is *magical* if it looks identical to a regular die and it is unfair; that is, the probabilities of obtaining $1, 2, \ldots, 6$ are not all identical. Two identical magical dice are thrown, and the numbers on the top faces are added. Determine whether it is possible that all possible sums (namely, $2, 3, \ldots, 12$) are equally likely.

8.8. [Euler's Identity] Let

$$f(x) = (1-x)(1-x^2)(1-x^3)\cdots = \prod_{n=1}^{\infty}(1-x^n).$$

Show that

$$[x^n]f(x) = \begin{cases} (-1)^k, & \text{if } n = \dfrac{3k^2 \pm k}{2}, \\ 0, & \text{otherwise.} \end{cases}$$

(The reader interested in variations of Euler's Identity, such as Gauss' Identity and the Gauss-Jacobi formula, are encouraged to read the papers by D.B. Fuchs and F.V. Vainshtein [61].)

8.9. Let ℓ be an even positive integer. Express

$$\sum_{k=0}^{n}\sum_{i=0}^{\ell}(-1)^i\binom{n}{k}^2\binom{2k}{i}\binom{2n-2k}{\ell-i}$$

in closed form.

8.10. [USAMO 1996, by Richard Stong] Determine whether there is a subset X of the integers with the following property: For any integer n there is exactly one solution of $a + 2b = n$ with $a, b \in X$.

8.11. Each vertex of a regular polygon is colored with one of a finite number of colors so that the points of the same color are the vertices of some new regular polygon. Prove that at least two of the polygons obtained are congruent.

8.12. Let $A_1, A_2, \ldots, B_1, B_2, \ldots$ be sets such that $A_1 = \emptyset$, $B_1 = \{0\}$,

$$A_{n+1} = \{x + 1 \mid x \in B_n\}, \quad \text{and} \quad B_{n+1} = A_n \cup B_n / A_n \cap B_n$$

for all positive integers n. Determine all n such that $B_n = \{0\}$.

9
Review Exercises

The problems marked with a * have also appeared in [12], which provides complete solutions to these problems.

9.1. [AHSME] An $11 \times 11 \times 11$ wooden cube is formed by gluing together 11^3 unit cubes. What is the greatest number of unit cubes that can be seen from a single point?

9.2. In how many ways may 1, 2, 3, 4, 5, and 6 be ordered so that no two consecutive terms have a sum that divisible by 2 or 3?

9.3. Find an integer n such that the decimal representation of 2003! ends in exactly n zeros.

9.4. A four-digit number $abcd$ is called *containing* if the number ad is bigger than the number bc. (For example, 4291 and 4091 are containing numbers, but 2359 is not a containing number.) Find the number of containing numbers.

9.5. [AIME 2003] Define a *good* word as a sequence of letters that consists only of the letters $A, B,$ and C (not all of these letters need to appear in the sequence), and in which A is never immediately followed by B, B is never immediately followed by C, and C is never immediately followed by A. How many seven-letter good words are there?

9.6. Let m and n be positive integers. Determine the number of sub-chessboards an $m \times n$ chessboard contains.

9.7. How many times is the word MATHEMATICS spelled along a path of Figure 9.1, where every two consecutive letters along the path must connect diagonally?

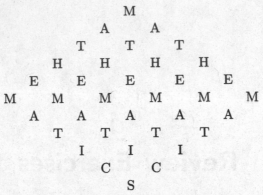

Figure 9.1.

9.8. [AMC12 2003] Let n be a five-digit number, and let q and r be the quotient and remainder, respectively, when n is divided by 100. For how many values of n is $q + r$ divisible by 11?

9.9. In how many ways can Mr. Fat travel from A_1 to A_{20} along the paths in Figure 9.2, assuming that he needs to go downward all of the time?

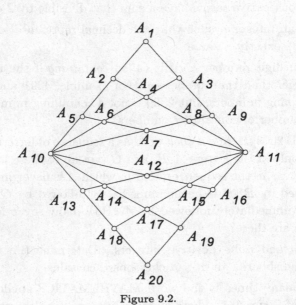

Figure 9.2.

9.10. [AIME 2000] A deck of forty cards consists of four 1's, four 2's, ... , and four 10's. A matching pair (two cards with the same number) is removed from the deck. Given that these cards are not returned to the deck, what is the probability that two randomly selected cards also form a pair?

9.11. Let n be a positive integer. One is given an $n \times 2$ chessboard and n dominoes. Determine the number of ways one can use the dominoes to tile the board.

9.12. [AMC12 2003] Let S be the set of permutations of the sequence 1, 2, 3, 4, 5 for which the first term is not 1. A permutation is chosen randomly from S. What is the probability that the second term is 2?

9.13. Find all possible values of n such that the decimal representation of $n!$ ends in exactly 2003 zeros.

9.14. Given that

$$\binom{n}{k} : \binom{n}{k+1} : \binom{n}{k+2} = 3 : 4 : 5,$$

find n.

9.15. [AHSME 1977] For how many paths consisting of a sequence of horizontal and/or vertical line segments, with each segment connecting a pair of adjacent letters in Figure 9.3, is the word CONTEST spelled out as the path is traversed?

C

COC

CONOC

CONTNOC

CONTETNOC

CONTESETNOC

CONTESTSETNOC

Figure 9.3.

9.16. Let n be a positive integer. Express

$$\sum_{i=0}^{n} \binom{n}{i} \sum_{j=i}^{n} \binom{n}{j}$$

in closed form.

9.17. Let n be a positive integer. An $n \times n$ array of numbers is *good* if all of the entries in the array are either $+1$ or -1, and the sum of the products of the entries in each row and column is equal to 0. Determine whether there exists a good array for $n = 4$ and $n = 5$.

9.18. Let $p(n, m, k)$ denote the number of m-part partitions of n with k as the smallest part. Prove that

$$p(n, m, k) = \begin{cases} p(n-1, m-1) & \text{if } k = 1, \\ p(n-m, m, k-1) & \text{if } k > 1. \end{cases}$$

9.19. Eight people are sitting in a row. In how many ways one can switch the seats of four of them, while the remaining four stay in their seats?

9.20. Determine the number of positive integers less than $1,000,000$ that contain only the digits 1, 2, 3, or 4.

9.21. How many squares have all four vertices in the set (Figure 9.4)

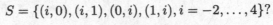
$$S = \{(i, 0), (i, 1), (0, i), (1, i), i = -2, \ldots, 4\}?$$

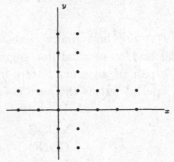

Figure 9.4.

9.22. [AMC12 2001] Given a regular nonagon, how many distinct equilateral triangles in the plane of the polygon have at least two vertices that are vertices of the nonagon?

9.23. Spider Fat is at vertex A of a cube. Fat spots a fly that stays at the point G on the cube that is furthest away from A. Assuming that Fat is always getting closer to the fly, in how many ways can Fat reach the fly by moving along the edges of the cube?

9.24. [AIME 1986] The polynomial

$$1 - x + x^2 - x^3 + \cdots + x^{16} - x^{17}$$

may be written in the form

$$a_0 + a_1y + a_2y^2 + \cdots + a_{16}y^{16} + a_{17}y^{17},$$

where $y = x + 1$ and the a_i are constants. Find a_2.

9.25. Let m and n be positive integers. Let $d(n, m)$ denote the number of partitions of n that have length m and are made up of distinct parts. Prove that

$$d(n, m) = d(n - m, m) + d(n - m, m - 1).$$

9.26. [AIME 1988] A convex polyhedron has for its faces 12 squares, 8 regular hexagons, and 6 regular octagons. At each vertex of the polyhedron one square, one hexagon, and one octagon meet. How many segments joining vertices of the polyhedron lie in the interior of the polyhedron, rather than along an edge or face?

9.27. At PEA, the chair of the Information Technology (IT) department has decided to set up 25 new servers in 13 academic buildings so that in each building there will be at least one new server. Outdoor cable connections is established between each pair of new servers that are in different buildings. Determine the least possible number of such connections.

9.28. In how many ways can Claudia put eight distinguishable balls into six distinct boxes, leaving exactly two boxes empty?

9.29. Determine the minimum value of n satisfying the following property: There is a sequence $a_1, a_2, \ldots, a_n$ such that for any subset S of the set $\{1, 2, 3, 4\}$, $S = \{a_i, a_{i+1}, \ldots, a_j\}$ for some $1 \le i \le j \le n$.

9.30. Amy, Ben, and Carol each sit at distinct vertices of an equilateral triangle. Diana, Emily, and Frank each sit at distinct midpoints of the sides of the same triangle. Each person wears either a red baseball cap or a blue baseball cap. In how many ways can they choose the colors of their baseball caps so that no three people wearing caps of the same color are sitting at points that determine an equilateral triangle? (Two coloring schemes are equivalent if one can be obtained from the other by rotation in the plane of the triangle.)

9.31. Let p be a prime. Prove that

$$\binom{n}{p} \equiv \left\lfloor \frac{n}{p} \right\rfloor \pmod{p}.$$

9.32.* [AIME 1992] A positive integer is called *ascending* if in its decimal representation there are at least two digits, and each digit is smaller than every digit to its right. How many ascending integers are there?

9.33. [MOSP 1998, by Franz Rothe] For a sequence $A_1, \ldots, A_n$ of subsets of $\{1, \ldots, n\}$ and a permutation π of $\{1, \ldots, n\}$, we define the diagonal set

$$D_\pi(A_1, \ldots, A_n) = \{i \in S : i \notin A_{\pi(i)}\}.$$

What is the maximum possible number of distinct sets that can occur as diagonal sets for a single choice of $A_1, \ldots, A_n$?

9.34. [AHSME 1995] For how many three-element sets of positive integers $\{a, b, c\}$ is it true that $a \times b \times c = 2310$?

9.35. [ARML 2002] Compute the number of ordered pairs (x, y) of integers with $1 \le x < y \le 100$ such that $i^x + i^y$ is a real number, where $i^2 = -1$.

9.36.* [AIME 1991] Given a rational number, write it as a fraction in lowest terms and calculate the product of the resulting numerator and denominator. For how many rational numbers between 0 and 1 will 20! be the resulting product?

9.37. Points $A_5, A_6, \ldots, A_{2003}$ are drawn on a sheet of paper inside the square $A_1 A_2 A_3 A_4$ such that no three distinct points A_i, A_j, and A_k are collinear. If Mr. Fat wants to cut the paper into triangular pieces so that each of the 2003 points becomes vertices of one of those pieces, how many cuts are required?

9.38.* [AIME 1985] In a tournament, each player played exactly one game against each of the other players. In each game the winner was awarded 1 point, the loser got 0 points, and each of the two players earned 1/2 point if the game was a tie. After the completion of the tournament, it was found that exactly half of the points earned by each player were earned in games against the ten players with the least number of points. (In particular, each of the ten lowest–scoring players earned half of his/her points against the other nine of the ten). What was the total number of players in the tournament?

9.39. Let T be a subset of $\{1, 2, \ldots, 2003\}$. An element a of T is called *isolated* if neither $a - 1$ nor $a + 1$ is in T. Determine the number of five-element subsets of T that contain no isolated elements.

9.40.* [AIME 1996] Two of the squares of a 7×7 checkerboard are painted yellow and the rest are painted green. Two color schemes are equivalent if one can be obtained from the other by applying a rotation in the plane of the board. How many inequivalent color schemes are possible?

9.41. Nine new teachers are going to be assigned to three offices, where each office can have at most four new teachers. In how many ways can this be done?

9.42. In how many ways can the set $\{1, 2, \ldots, 2003\}$ be partitioned into three nonempty sets such that none of these sets contain a pair of consecutive integers?

9.43. Find all integers n such that $1 \leq n \leq 2003$ and the decimal representation of $m!$ ends in exactly n zeros for some positive integer m.

9.44. [AIME 1991] Two three-letter strings, aaa and bbb, are transmitted electronically. Each string is sent letter by letter. Due to faulty equipment, each of the six letters has a $\frac{1}{3}$ chance of being received incorrectly (as an a when its should have been a b, or as a b when it should have been an a). However, whether a given letter is received correctly or incorrectly is independent of the reception of any other letter. Let S_a be the three-letter string received when aaa is transmitted, and let S_b be the three-letter string received when bbb is transmitted. What is the probability that S_a comes before S_b in alphabetical order?

9.45. [China 1990] In how many ways can 8 girls and 25 boys be arranged around a circular table so that there are at least two boys between any two girls?

9.46. [AIME 2003] An integer between 1000 and 9999, inclusive, is called *balanced* if the sum of its leftmost two digits equals the sum of its rightmost two digits. How many balanced integers are there?

9.47. Let n be a positive integer and let X be a set with n elements.

Express

$$\sum_{A,B \subseteq X} |A \cap B|$$

in closed form.

9.48. [AIME 1992] Let S be the set of all rational numbers r, $0 < r < 1$, that have a repeating decimal expansion of the form

$$0.abcabcabc \cdots = 0.\overline{abc},$$

where the digits a, b, and c are not necessarily distinct. To write the elements of S as fractions in lowest terms, how many different numerators are required?

9.49. [Vietnam 1996] Determine, as a function of n, the number of permutations of the set $\{1, 2, \cdots, n\}$ such that no three of the integers 1, 2, 3, and 4 appear consecutively.

9.50. [Canada 1991] Ten distinct numbers from the set $\{0, 1, 2, \ldots, 14\}$ are chosen to fill the ten circles shown in Figure 9.5. The absolute value of the difference of the two numbers joined by each segment must be different for different segments. Is it possible to do this? Justify your answer.

Figure 9.5.

9.51.* [AIME 1996] A bored student walks down a hall that contains a row of closed lockers numbered 1 to 1024. He opens the locker numbered 1, and then alternates between skipping and opening each closed locker thereafter. When he reaches the end of the hall, the student turns around opens the first closed locker he encounters, and then alternates between skipping and opening the closed lockers. The student continues wandering back and forth in this manner until every locker is open. What is the number of the last locker he opens?

9.52. A 2002×2004 chessboard is given with a 0 or 1 written in each square so that each row and column contain an odd number of squares containing a 1. Prove that there is an even number of white unit squares containing 1.

9.53. [AIME 2003] What is the number of positive integers that are less than or equal to 2003 and whose binary (base 2) representations have more 1's than 0's?

9.54. A triangulation of a polygon is a partitioning of the polygon into triangles all of whose vertices are vertices of the original polygon. Given an n-side convex polygon, determine the number of different possible triangulations of it.

9.55. Show that Examples 4.11 and 8.4 are equivalent to each other.

9.56. [ARML 2003] Let N be a three-digit number that is divisible by three. One of N's digits is chosen at random and is removed. Compute the probability that the remaining number is divisible by three. (Note that if a digit is removed from 207 we obtain either 20, 07 = 7, or 27; if a digit is removed from 300, we obtain either 30, 30, or 00 = 0.)

9.57. [AIME 1994] A solitaire game is played as follows. Six distinct pairs of matched tiles are placed in a bag. The player randomly draws tiles one at a time from the bag and keeps them, except that matching tiles are put aside as soon as they appear in the player's hand. The game ends if the player ever holds three tiles, no two of which match; otherwise, the drawing continues until the bag is empty. Compute the probability that the bag will be emptied.

9.58. [China 1989] Let n be a positive integer with $n \geq 4$. Each entry in an $n \times n$ matrix is either 1 or −1. A *base* is defined to be the product of n entries, no two of which are in the same row or column. Prove that the sum of all the bases is divisible by 4.

9.59.* [AIME 1997] How many different 4×4 arrays whose entries are all 1's and −'s, have the property that the sum of the entries in each row and column is 0?

9.60. Let p be a prime and let n be an integer with $n \geq p$. Prove that p divides

$$\sum_{k=0}^{\lfloor \frac{n}{p} \rfloor} (-1)^k \binom{n}{pk}.$$

9.61. [AIME 1999] Ten points in the plane are given, with no three collinear. Four distinct segments joining pairs of these points are chosen at random, all such segments being equally likely. What is the probability that some three of the segments form a triangle whose vertices are among the ten given points?

9.62. Let n and m be positive integers with $m \leq n$. Determine the number of sequences of integers $\{a_k\}_{k=1}^m$ such that $1 \leq a_1 < a_2 < \cdots < a_m \leq n$, and for $1 \leq i \leq m$, a_i and i have the same parity; that is, $a_i - i$ is even.

9.63. Let n be a nonnegative integer. Express

$$\sum_{k=0}^n \frac{1}{(n-k)!(n+k)!}$$

in closed form.

9.64. [AIME 1999] Forty teams play a tournament in which every team plays every other team exactly once. No ties occur, and each team has a 50% chance of winning any game that it plays. What is the probability that no two teams win the same number of games?

9.65. [Hungary 1998] Two players take turns drawing, with replacement, a card at random from a deck of four cards labeled 1, 2, 3, and 4. The game stops as soon as the sum of the numbers that have appeared since the start of the game is divisible by 3, and the player who drew the last card is the winner. What is the probability that the player who goes first wins?

9.66. [IMO 2002] Let n be an odd integer greater than 1 and let $c_1, c_2, \ldots, c_n$ be integers. For each permutation $a = (a_1, a_2, \ldots, a_n)$ of $\{1, 2, \ldots, n\}$, define $S(a) = \sum_{i=1}^n c_i a_i$. Prove that there exist permutations b and c, $b \neq c$, such that $n!$ divides $S(b) - S(c)$.

9.67.* [AIME 1986] In a sequence of coin tosses, one can keep a record of the number of instances in which a tail is immediately followed by a head, a head is immediately followed by a head, etc. We denote these by TH, HH, etc. For example, in the sequence $HHTTHHHHTHHTTTT$ of 15 coin tosses, we observe that there are five HH, three HT, two TH, and four TT subsequences. How many different sequences of 15 coin tosses will contain exactly two HH, three HT, four TH, and five TT subsequences?

9.68. Let n be a positive integer. Determine the number of 3×3

matrices

$$\begin{bmatrix} a_1 & b_1 & c_1 \\ a_2 & b_2 & c_2 \\ a_3 & b_3 & c_3 \end{bmatrix}$$

such that $a_i + b_i + c_i = n$, $a_i \geq 0$, $b_i \geq 0$, and $c_i \geq 0$ for $i = 1, 2, 3$.

9.69. [AIME 2001] Club Truncator is in a soccer league with six other teams, each of which it plays once. In any of its 6 matches, the probabilities that Club Truncator will win, lose, or tie are each 1/3. What is the probability that Club Truncator will finish the season with more wins than losses?

9.70. Let n be a positive integer. We are given an $n \times n$ chessboard. We call the intersections of k adjacent rows and ℓ adjacent columns a $k \times \ell$ *sub-board*, and say that $k + \ell$ is the semiperimeter of the sub-board. Assume that some sub-boards with semiperimeters of n cover the **main diagonal** (the diagonal from the top left to the bottom right) of the board. Prove that the sub-boards cover at least half of the unit squares of the board.

9.71. Let k, m, and n be positive integers. There are n distinguishable chairs evenly placed around a circular table. There are k students to be seated around the table in such a way that there are at least m empty chairs between each pair of students. In how many ways can this be done?

9.72. Let n be the number of incongruent triangles such that their side lengths are all integers and the sums of the lengths of two of their sides are equal to $2m + 1$ for positive integers m. Find n in terms of m.

9.73. [AIME 1993] Alfred and Bonnie play a game in which they take turns tossing a fair coin. The winner of a game is the first person to obtain a head. Alfred and Bonnie play this game several times, with the stipulation that the loser of each game goes first in the next game. Suppose Alfred goes first in the first game. What is the probability that he will win the sixth game?

9.74. [APMO 1998] Let F be the set of all n-tuples $(A_1, \ldots, A_n)$ such that each A_i is a subset of $\{1, 2, \ldots, 1998\}$. Express

$$\sum_{(A_1, \ldots, A_n) \in F} \left| \bigcup_{i=1}^{n} A_i \right|$$

in closed form.

9.75. Let n be a positive integer. Prove that

$$\sum_{k=0}^{n} \frac{1}{k+1}\binom{n}{k} = \frac{1}{n+1}(2^{n+1} - 1).$$

9.76. [IMO 1985 Short–listed] Let n be a positive integer, and let $P_1, P_2, \ldots, P_n$ be distinct two-element subsets of the set

$$S = \{a_1, a_2, \ldots, a_n\}$$

such that if $P_i \cap P_j \neq \emptyset$, then $\{a_i, a_j\}$ is one of the P's. Prove that each of the a's appears in exactly two of the P's.

9.77.* [AIME 1992] In the game of *Chomp*, two players alternately take "bites" from a 5-by-7 grid of unit squares. To take a bite, the player chooses one of the remaining squares, then removes ("eats") all squares found in the quadrant defined by the left edge (extended upward) and the lower edge (extended rightward) of the chosen square. For example, the bite determined by the shaded square in Figure 9.5 would remove the shaded square and the four squares marked by ×'s.

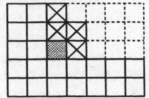

Figure 9.5.

(The squares with two or more dotted edges were removed from the original board on previous moves.) The object of the game is to make one's opponent take the last bite. The diagram shows one of the many subsets of the set of 35 unit squares that can occur during the game of Chomp. How many different subsets are there in all? Include the full board and the empty board in your count.

9.78. Fifteen points $a_1, a_2, \ldots, a_{15}$ are evenly distributed around a circle. A pair of points a_i and a_j is called a *combo* if there are two points on the minor arc $\widehat{a_i a_j}$. Determine the number of ways to pick sets of points without a combo.

9.79. [AIME 1996] For each permutation $a_1, a_2, a_3, \ldots, a_{10}$ of integers $1, 2, 3, \ldots, 10$, form the sum

$$|a_1 - a_2| + |a_3 - a_4| + |a_5 - a_6| + |a_7 - a_8| + |a_9 - a_{10}|.$$

What is the average value of all such sums?

9.80. Let m and n be positive integers with $m \leq n$. Express

$$\sum_{k=m}^{n} (-1)^{k+m} \binom{n}{k} \binom{k}{m}$$

in closed form.

9.81. Let k and n be positive integers, and let $a_1, a_2, \ldots, a_k$ be positive integers. Determine the number of ways to choose a $(k + 1)$-term sequence $\{b_i\}_{i=1}^{k+1}$ from the first n positive integers such that $b_{i+1} - b_i \geq a_i$ for $1 \leq i \leq k - 1$.

9.82. Let A and B be disjoint finite sets of integers, and let a and b be positive integers such that if $x \in A \cap B$, then either $x + a \in A$ or $x - b \in B$. Prove that $a|A| = b|B|$.

9.83. Let n be a positive integer. Show that

$$\sum_{k=0}^{n} \binom{2n}{k} = 2^{2n-1} + \frac{1}{2} \binom{2n}{n}.$$

9.84.* [AIME 1988] In an office, at various times during the day the boss gives the secretary a letter to type, each time putting the letter on the top of the pile in the secretary's in-box. When there is time, the secretary takes the top letter off the pile and types it. There are nine letters to be typed during the day, and the boss delivers them in the order $1, 2, 3, 4, 5, 6, 7, 8, 9$. While leaving for lunch, the secretary tells a colleague that letter 8 has already been typed, but says nothing else about the rest of the morning's typing. The colleague wonders which of the nine letters remain to be typed after lunch and in what order they will be typed. Based upon the above information, how many such *after-lunch typing orders* are possible? (One possibility is that there are no letters left to be typed.)

9.85. [Bulgaria 1997, by Ivan Landgev] Let n be a positive integer. Find the number of $2n$-digit positive integers $a_1 a_2 \ldots a_{2n}$ such that

(i) none of the digits a_i is equal to 0, and

(ii) the sum $a_1a_2 + a_3a_4 + \cdots + a_{2n-1}a_{2n}$ is even.

9.86. Let k and n be positive integers, and let $a_1, a_2, \ldots, a_k$ be positive integers. Marbles $m_1, m_2, \ldots, m_n$ are placed around a circle in clockwise order. Zachary is asked to pick k marbles $m_{j_1}, m_{j_2}, \ldots, m_{j_k}$, with $j_1 < j_2 < \cdots < j_k$, such that there are at least a_i marbles that are not picked between m_{j_i} and $m_{j_{i+1}}$ in the clockwise direction. (Here marble m_{n+k} is the same as marble m_k.) In how many ways can Zachary complete his task?

9.87. Let n be a positive integer. Prove that

$$\sum_{k=0}^{\lfloor \frac{n}{2} \rfloor} \left[\binom{n}{k} - \binom{n}{k-1} \right]^2 = \frac{1}{n+1}\binom{2n}{n}.$$

9.88.* [MOSP 2001] A class of fifteen boys and fifteen girls is seated around a round table. Their teacher wishes to pair up the students and hand out fifteen tests, one test to each pair. As the teacher is preparing to select the pairs to hand out the tests, he wonders to himself, "How many seating arrangements would allow me to match up boy/girl pairs sitting next to each other without having to ask any student to change his or her seat?" Answer the teacher's question. (Two seating arrangements are regarded as being the same if one can be obtained from the other by rotation.)

9.89. Prove that

$$\sum(a_1a_2 \cdots a_n) = \prod_{k=1}^{n}(2k-1) = (2k-1)!!,$$

where the sum is taken over all ordered n-tuples of integers $(a_1, a_2, \ldots, a_n)$ with $1 \le a_i \le n+1-i$ and $a_{i+1} \ge a_i - 1$ for all i.

9.90. Let n be a positive integer. A convex polygon $A_1A_2 \ldots A_n$ is inscribed in the circle ω. All of the diagonals of the polygon are drawn. Determine the maximum number of regions that are enclosed by the sides and diagonals of the polygon and the arcs of ω.

9.91. [China 1999] Let n be a positive integer, and let $S = \{1, 2, \ldots, n\}$. A 4-*double cover* of S is a 4-element set of subsets of S such that each element of S belongs to exactly two subsets. Determine the number of 4-double covers of S.

9.92. Each edge of a given polyhedron $\mathcal{P}$ has been colored either red or blue. We say that a face angle is *singular* if the two edges forming the angle have different colors. For each vertex P_i of $\mathcal{P}$, let S_{P_i} denote the number of singular face angles with P_i as their vertex. Prove that there exist vertices P_i and P_j with $S_{P_i} + S_{P_j} \leq 4$.

9.93.* [AIME 1998] Define a *domino* to be an ordered pair of distinct positive integers. A *proper sequence* of dominoes is a list of distinct dominoes in which the first coordinate of each pair after the first equals the second coordinate of the immediately preceding pair, and in which (i, j) and (j, i) do not both appear for any i and j. Let D_{40} be the set of all dominoes whose coordinates are no larger than 40. Find the length of the longest proper sequence of dominoes that can be formed using the dominoes of D_{40}.

9.94. Let m and n be positive integers, and let

$$S(m,n) = \binom{n}{0}(2^n - 1)^m - \binom{n}{1}(2^{n-1} - 1)^m + \binom{n}{2}(2^{n-2} - 1)^m$$

$$- \cdots + (-1)^{n-1}\binom{n}{n-1}.$$

Prove that $S(m,n) = S(n,m)$.

9.95.* [AIME 2000] A stack of 2000 cards is labeled with the integers from 1 to 2000, with different integers on different cards. The cards in the stack are not in numerical order. The top card is removed from the stack and placed on the table, and the next card in the stack is moved to the bottom of the stack. The new top card is removed from the stack and placed on the table to the right of the card already there, and the next card in the stack is moved to the bottom of the stack. This process—placing the top card to the right of the cards already on the table and moving the next card in the stack to the bottom of the stack—is repeated until all of the cards are on the table. It is found that, reading left to right, the labels of the cards on the table are in ascending order: $1, 2, 3, \ldots, 1999, 2000$. In the original stack of cards, how many cards were above the card labeled 1999?

9.96. Let n and k be positive integers such that there exist nonnegative integers $n_1, n_2, \ldots, n_k$ with

$$n_1 + 2n_2 + \cdots + kn_k = n.$$

Let S be a set with n elements. Determine the number of partitions of S such that for $1 \le i \le k$, have exactly n_i subsets with i elements.

9.97. Let $p_n(k)$ be the number of permutations of the set $\{1, \ldots, n\}$, with $n \ge 1$, that have exactly k fixed points. Prove that

$$\sum_{k=0}^{n} k p_n(k) = \sum_{k=0}^{n} (k-1)^2 p_n(k).$$

9.98.* [AIME 2001] A mail carrier delivers mail to the nineteen houses on the east side of Elm Street. The carrier notices that no two adjacent houses ever get mail on the same day, but that there are never more than two houses in a row that get no mail on the same day. How many different patterns of mail delivery are possible?

9.99. [China 1999, by Zhusheng Zhang] A $4 \times 4 \times 4$ cube is composed of 64 unit cubes. The faces of 16 unit cubes are colored red. A coloring is called *interesting* if there is exactly 1 red unit cube in every $1 \times 1 \times 4$ rectangular box of 4 unit cubes. Determine the number of interesting colorings. (Two colorings are different even if one can be transformed into another by a series of rotations.)

9.100. Let m be a positive integer. Determine the number of sets $\{a, b, c, d\} \subset \{1, 2, \ldots, 2m\}$ such that $a + b = c + d$.

9.101. [Cauchy's Identity] Let n and r be positive integers. Prove that

$$\sum_{\substack{n_1 + 2n_2 + \cdots + rn_r = n \\ n_1, n_2, \ldots, n_r \ge 0}} \frac{1}{n_1! n_2! \cdots n_r! 1^{n_1} 2^{n_2} \cdots r^{n_r}} = 1.$$

9.102. One is given a 2×12 chessboard and an ample supply of 1×2 and 1×3 card boards. Numbers $1, 2, \ldots, 12$ and $13, 14, \ldots, 24$ are written in each unit square, in that order from left to right, in the top and bottom row, respectively. In how many ways can the chessboard be tiled by a collection of card boards?

9.103. [MOSP 2001, from Kvant] An alphabet consists of three letters. Some sequences of letters of length two or more are prohibited, and all of the prohibited sequences have different lengths. A *word* is a sequence of letters of any length. A *correct* word does not contain any prohibited sequences. Prove that there are correct words of any length.

9.104. [Balkan 1990] Let $\mathcal{F}$ be a set of subsets of $\{1, 2, \ldots, n\}$ such that

   (i) every element in $\mathcal{F}$ is a three-element subset; and

   (ii) the intersection of any two elements of $\mathcal{F}$ contains at most one element.

   Let f denote the maximum number of elements that $\mathcal{F}$ can have. Prove that $n^2 - 4n \le 6f \le n^2 - n$.

9.105. Let p be a prime with $p \ge 5$. Prove that

$$\binom{2p}{p} \equiv 2 \pmod{p^3}.$$

9.106. [USAMO 1984] A difficult mathematical competition consisted of Part I and Part II, with a combined total of 28 problems. Each contestant solved exactly 7 problems altogether. For each pair of problems, there were exactly two contestants who solved both of them. Prove that there was a contestant who solved either no problems or at least 4 problems in Part I.

9.107. [China 2000, by Chengzhang Li] An exam paper consists of five multiple-choice questions, each with four different choices; 2000 students take the test, and each student chooses exactly one answer per question. Find the smallest value of n for which it is possible for the students' answer sheets have the following property: among any n of the students' answer sheets, there exist four of them among which any two have at most three common answers.

9.108. [USAMO 2001, by Kiran Kedlaya] Each of eight boxes contains six balls. Each ball has been colored with one of n colors, such that no two balls in the same box are the same color, and no two colors occur together in more than one box. Determine, with justification, the smallest integer n for which this is possible. (Let m be a positive integer. What if there are $m + 2$ boxes and n balls?)

9.109. Let m be a positive integer. In the coordinate plane, we call a polygon $A_1 A_2 \ldots A_n$ *admissible* if

   (i) its perimeter $A_1 A_2 + A_2 A_3 + \cdots + A_n A_1$ is equal to m;

   (ii) for $1 \le i \le n$, $A_i = (x_i, y_i)$ is a lattice point;

     (iii) $x_1 \leq x_2 \leq \cdots \leq x_n$; and

     (iv) $y_1 = y_n = 0$, and there is a k with $1 < k < m$ such that $y_1 \leq y_2 \cdots \leq y_k$ and $y_k \geq y_{k+1} \geq \cdots \geq y_n$.

Determine the number of admissible polygons as a function of m. (Two different admissible polygons are considered distinct even if one can be obtained from the other via composition of reflections and rotations.)

     (This problem is due to David Vincent, who studied this topic at the Clay Mathematics Spring Program in March 2003. Tiankai Liu and David developed two wonderful solutions on the PEA math team's bus trip to the ARML meet, May 31, 2003.)

9.110. [IMC 1998] Let $A_n = \{1, 2, \ldots, n\}$, where n is a positive integer greater then two. Let F_n be the family of all inconstant functions $f : A_n \to A_n$ satisfying the following conditions:

     (i) $f(k) \leq f(k+1)$ for $k = 1, 2, \ldots, n-1$; and

     (ii) $f(k) = f(f(k+1))$ for $k = 1, 2, \ldots, n-1$.

Find the number of functions in F_n.

9.111.* [China 1999, by Hongbin Yu] For a set S, let $|S|$ denote the number of elements in S. Let A be a set with $|A| = n$, and let $A_1, A_2, \ldots, A_n$ be subsets of A with $|A_i| \geq 2$ and $1 \leq i \leq n$. Suppose that for each two-element subset A' of A there is a unique i such that $A' \subseteq A_i$. Prove that $A_i \cap A_j \neq \emptyset$ for $1 \leq i < j \leq n$.

Glossary

Convexity A function $f(x)$ is **concave up (down)** on $[a, b] \subseteq \mathbb{R}$ if $f(x)$ lies under (over) the line connecting $(a_1, f(a_1))$ and $(b_1, f(b_1))$ for all

$$a \leq a_1 < x < b_1 \leq b.$$

A function $g(x)$ is concave up (down) on the Euclidean plane if it is concave up (down) on each line in the plane, where we identify the line naturally with $\mathbb{R}$.

Concave up and down functions are also called **convex** and **concave**, respectively.

If f is concave up on an interval $[a, b]$ and $\lambda_1, \lambda_2, \ldots, \lambda_n$ are nonnegative numbers with sum equal to 1, then

$$\lambda_1 f(x_1) + \lambda_2 f(x_2) + \cdots + \lambda_n f(x_n) \geq f(\lambda_1 x_1 + \lambda_2 x_2 + \cdots + \lambda_n x_n)$$

for any $x_1, x_2, \ldots, x_n$ in the interval $[a, b]$. If the function is concave down, the inequality is reversed. This is **Jensen's Inequality**.

Lagrange's Interpolation Formula Let $x_0, x_1, \ldots, x_n$ be distinct real numbers, and let $y_0, y_1, \ldots, y_n$ be arbitrary real numbers. Then there exists a unique polynomial $P(x)$ of degree at most n such that $P(x_i) = y_i$, $i = 0, 1, \ldots, n$. This polynomial is given by

$$P(x) = \sum_{i=0}^{n} y_i \frac{(x - x_0) \cdots (x - x_{i-1})(x - x_{i+1}) \cdots (x - x_n)}{(x_i - x_0) \cdots (x_i - x_{i-1})(x_i - x_{i+1}) \cdots (x_i - x_n)}.$$

Maclaurin Series Given a function $f(x)$, the power series

$$\sum_{k=0}^{\infty} \frac{f^{(k)}(0)}{k!} x^k$$

213

is the **Maclaurin Series** of $f(x)$, where $f^{(k)}(x)$ denotes the kth derivative of $f(x)$.

Pigeonhole Principle If n objects are distributed among $k < n$ boxes, some box contains at least two objects.

Power Mean Inequality Let $a_1, a_2, \ldots, a_n$ be any positive numbers for which $a_1 + a_2 + \cdots + a_n = 1$. For positive numbers $x_1, x_2, \ldots, x_n$ we define

$$M_{-\infty} = \min\{x_1, x_2, \ldots, x_k\},$$

$$M_{\infty} = \max\{x_1, x_2, \ldots, x_k\},$$

$$M_0 = x_1^{a_1} x_2^{a_2} \cdots x_n^{a_n},$$

$$M_t = (a_1 x_1^t + a_2 x_2^t + \cdots + a_k x_k^t)^{1/t},$$

where t is a non-zero real number. Then

$$M_{-\infty} \leq M_s \leq M_t \leq M_{\infty}$$

for $s \leq t$.

Root Mean Square-Arithmetic Mean Inequality For positive numbers $x_1, x_2, \ldots, x_n$,

$$\sqrt{\frac{x_1^2 + x_2^2 + \cdots + x_k^2}{n}} \geq \frac{x_1 + x_2 + \cdots + x_k}{n}.$$

The inequality is a special case of the **Power Mean Inequality**.

Triangle Inequality In a non-degenerated triangle, the sum of the lengths of any two sides of the triangle is bigger than the length of the third side.

Vandermonde Matrix A **Vandermonde Matrix** $\mathbf{M}$ is a matrix of the form

$$\mathbf{M} = \begin{bmatrix} 1 & 1 & \cdots & 1 \\ x_1 & x_2 & \cdots & x_n \\ x_1^2 & x_2^2 & \cdots & x_n^2 \\ \cdots & \cdots & \cdots & \cdots \\ x_1^{n-1} & x_2^{n-1} & \cdots & x_n^{n-1} \end{bmatrix}.$$

Its determinant is

$$\prod_{1 \leq i < j \leq n} (x_j - x_i),$$

which is nonzero if and only if $x_1, x_2, \ldots, x_n$ are distinct. The Vandermonde matrix is closely related to the **Lagrange's Interpolation Formula**. Indeed, it arises in the problem of finding a polynomial

$$p(x) = a_{n-1}x^{n-1} + a_{n-2}x^{n-2} + \cdots + a_1 x + a_0$$

such that $p(x_i) = y_i$ for all i with $1 \leq i \leq n$. Because

$$a_{n-1}x_1^{n-1} + a_{n-2}x_1^{n-2} + \cdots + a_1 x_1 + a_0 = y_1,$$

$$a_{n-1}x_2^{n-1} + a_{n-2}x_2^{n-2} + \cdots + a_1 x_2 + a_0 = y_2,$$

$$\ldots\ldots\ldots$$

$$a_{n-1}x_n^{n-1} + a_{n-2}x_2^{n-2} + \cdots + a_1 x_n + a_0 = y_n,$$

it follows that

$$\begin{bmatrix} y_1 \\ y_2 \\ \cdots \\ y_n \end{bmatrix} = \mathbf{M}^T \cdot \begin{bmatrix} a_0 \\ a_1 \\ \cdots \\ a_{n-1} \end{bmatrix}$$

$$= \begin{bmatrix} 1 & x_1 & x_2^2 & \cdots & x_1^{n-1} \\ 1 & x_2 & x_2^2 & \cdots & x_2^{n-1} \\ \cdots & \cdots & \cdots & \cdots & \cdots \\ 1 & x_n & x_n^2 & \cdots & x_n^{n-1} \end{bmatrix} \begin{bmatrix} a_0 \\ a_1 \\ \cdots \\ a_{n-1} \end{bmatrix},$$

where $\mathbf{M}^T$ is the transpose of $\mathbf{M}$. (Note that a matrix and its transpose have the same determinant.)

Index

Further Reading

1. Andreescu, T.; Feng, Z., *USA and International Mathematical Olympiads 2002* , Mathematical Association of America, 2003.

2. Andreescu, T.; Feng, Z., *USA and International Mathematical Olympiads 2001* , Mathematical Association of America, 2002.

3. Andreescu, T.; Feng, Z., *USA and International Mathematical Olympiads 2000* , Mathematical Association of America, 2001.

4. Andreescu, T.; Feng, Z.; Lee, G.; Loh, P., *Mathematical Olympiads: Problems and Solutions from around the World, 2001–2002*, Mathematical Association of America, 2004.

5. Andreescu, T.; Feng, Z.; Lee, G., *Mathematical Olympiads: Problems and Solutions from around the World, 2000–2001*, Mathematical Association of America, 2003.

6. Andreescu, T.; Feng, Z., *Mathematical Olympiads: Problems and Solutions from around the World, 1999–2000*, Mathematical Association of America, 2002.

7. Andreescu, T.; Feng, Z., *Mathematical Olympiads: Problems and Solutions from around the World, 1998–1999*, Mathematical Association of America, 2000.

8. Andreescu, T.; Kedlaya, K., *Mathematical Contests 1997–1998: Olympiad Problems from around the World, with Solutions*, American Mathematics Competitions, 1999.

9. Andreescu, T.; Kedlaya, K., *Mathematical Contests 1996–1997: Olympiad Problems from around the World, with Solutions*, American Mathematics Competitions, 1998.

10. Andreescu, T.; Kedlaya, K.; Zeitz, P., *Mathematical Contests*

1995–1996: Olympiad Problems from around the World, with Solutions, American Mathematics Competitions, 1997.

11. Andreescu, T.; Feng, Z., *101 Problems in Algebra from the Training of the USA IMO Team*, Australian Mathematics Trust, 2001.

12. Andreescu, T.; Feng, Z., *102 Combinatorial Problems from the Training of the USA IMO Team*, Birkhäuser, 2002.

13. Andreescu, T.; Enescu, B., *Mathematical Olympiad Treasures*, Birkhäuser, 2003.

14. Andreescu, T.; Gelca, R., *Mathematical Olympiad Challenges*, Birkhäuser, 2000.

15. Andreescu, T.; Andrica, D., *360 Problems for Mathematical Contests*, GIL, 2002.

16. Andreescu, T.; Andrica, D., *An Introduction to Diophantine Equations*, GIL, 2002.

17. Barbeau, E., *Polynomials*, Springer-Verlag, 1989.

18. Beckenbach, E. F.; Bellman, R., *An Introduction to Inequalities*, New Mathematical Library, Vol. 3, Mathematical Association of America, 1961.

19. Bollobas, B., *Graph Theory, An Introductory Course*, Springer-Verlag, 1979.

20. Chinn, W. G.; Steenrod, N. E., *First Concepts of Topology*, New Mathematical Library, Vol. 27, Random House, 1966.

21. Cofman, J., *What to Solve?*, Oxford Science Publications, 1990.

22. Coxeter, H. S. M.; Greitzer, S. L., *Geometry Revisited*, New Mathematical Library, Vol. 19, Mathematical Association of America, 1967.

23. Coxeter, H. S. M., *Non-Euclidean Geometry*, The Mathematical Association of American, 1998.

24. Doob, M., *The Canadian Mathematical Olympiad 1969–1993*, University of Toronto Press, 1993.

25. Engel, A., *Problem-Solving Strategies*, Problem Books in Mathematics, Springer, 1998.

26. Fomin, D.; Kirichenko, A., *Leningrad Mathematical Olympiads 1987–1991*, MathPro Press, 1994.

27. Fomin, D.; Genkin, S.; Itenberg, I., *Mathematical Circles*, American Mathematical Society, 1996.

28. Graham, R. L.; Knuth, D. E.; Patashnik, O., *Concrete Mathematics*, Addison-Wesley, 1989.

29. Gillman, R., *A Friendly Mathematics Competition*, The Mathematical Association of American, 2003.

30. Greitzer, S. L., *International Mathematical Olympiads, 1959–1977*, New Mathematical Library, Vol. 27, Mathematical Association of America, 1978.

31. Grossman, I.; Magnus, W., *Groups and Their Graphs*, New Mathematical Library, Vol. 14, Mathematical Association of America, 1964.

32. Holton, D., *Let's Solve Some Math Problems*, A Canadian Mathematics Competition Publication, 1993.

33. Ireland, K.; Rosen, M., *A Classical Introduction to Modern Number Theory*, Springer-Verlag, 1982.

34. Kazarinoff, N. D., *Geometric Inequalities*, New Mathematical Library, Vol. 4, Random House, 1961.

35. Kedlaya, K; Poonen, B.; Vakil, R., *The William Lowell Putnam Mathematical Competition 1985–2000*, The Mathematical Association of American, 2002.

36. Klamkin, M., *International Mathematical Olympiads, 1978–1985*, New Mathematical Library, Vol. 31, Mathematical Association of America, 1986.

37. Klamkin, M., *USA Mathematical Olympiads, 1972–1986*, New Mathematical Library, Vol. 33, Mathematical Association of America, 1988.

38. Klee, V.; Wagon, S, *Old and New Unsolved Problems in Plane Geometry and Number Theory*, The Mathematical Association of American, 1991.

39. Kürschák, J., *Hungarian Problem Book, volumes I & II*, New Mathematical Library, Vols. 11 & 12, Mathematical Association of America, 1967.

40. Kuczma, M., *144 Problems of the Austrian–Polish Mathematics Competition 1978–1993*, The Academic Distribution Center, 1994.

41. Landau, E., *Elementary Number Theory*, Chelsea Publishing Company, New York, 1966.

42. Larson, L. C., *Problem-Solving Through Problems*, Springer-Verlag, 1983.

43. Lausch, H. *The Asian Pacific Mathematics Olympiad 1989–1993*, Australian Mathematics Trust, 1994.

44. Leveque, W. J., *Topics in Number Theory, Volume 1*, Addison Wesley, New York, 1956.

45. Liu, A., *Chinese Mathematics Competitions and Olympiads 1981–1993*, Australian Mathematics Trust, 1998.

46. Liu, A., *Hungarian Problem Book III*, New Mathematical Library, Vol. 42, Mathematical Association of America, 2001.

47. Lozansky, E.; Rousseau, C. *Winning Solutions*, Springer, 1996.

48. Mordell, L. J., *Diophantine Equations*, Academic Press, London and New York, 1969.

49. Ore, O., *Graphs and Their Use, Random House*, 1963.

50. Ore, O., *Invitation to Number Theory, Random House*, 1967.

51. Savchev, S.; Andreescu, T. *Mathematical Miniatures*, Anneli Lax New Mathematical Library, Vol. 43, Mathematical Association of American, 2002.

52. Sharygin, I. F., *Problems in Plane Geometry*, Mir, Moscow, 1988.

53. Sharygin, I. F., *Problems in Solid Geometry*, Mir, Moscow, 1986.

54. Shklarsky, D. O; Chentzov, N. N; Yaglom, I. M., *The USSR Olympiad Problem Book*, Freeman, 1962.

55. Slinko, A., *USSR Mathematical Olympiads 1989–1992*, Australian Mathematics Trust, 1997.

56. Sierpinski, W., *Elementary Theory of Numbers*, Hafner, New York, 1964.

57. Soifer, A., *Colorado Mathematical Olympiad: The first ten years*, Center for excellence in mathematics education, 1994.

58. Szekely, G. J., *Contests in Higher Mathematics*, Springer-Verlag, 1996.

59. Stanley, R. P., *Enumerative Combinatorics*, Cambridge University Press, 1997.

60. Tabachnikov, S. *Kavant Selecta: Algebra and Analysis I*, American Mathematics Society, 1991.

61. Tabachnikov, S. *Kavant Selecta: Algebra and Analysis II*, American Mathematics Society, 1991.

62. Tabachnikov, S. *Kavant Selecta: Combinatorics I*, American Mathematics Society, 2000.

63. Taylor, P. J., *Tournament of Towns 1980–1984*, Australian Mathematics Trust, 1993.

64. Taylor, P. J., *Tournament of Towns 1984–1989*, Australian Mathematics Trust, 1992.

65. Taylor, P. J., *Tournament of Towns 1989–1993*, Australian Mathematics Trust, 1994.

66. Taylor, P. J.; Storozhev, A., *Tournament of Towns 1993–1997*, Australian Mathematics Trust, 1998.

67. Tomescu, I., *Problems in Combinatorics and Graph Theory*, Wiley, 1985.

68. Vanden Eynden, C., *Elementary Number Theory*, McGraw-Hill, 1987.

69. Vaderlind, P.; Guy, R.; Larson, L., *The Inquisitive Problem Solver*, The Mathematical Association of American, 2002.

70. Wilf, H. S., *Generatingfunctionology*, Academic Press, 1994.

71. Wilson, R., *Introduction to graph theory*, Academic Press, 1972.

72. Yaglom, I. M., *Geometric Transformations*, New Mathematical Library, Vol. 8, Random House, 1962.

73. Yaglom, I. M., *Geometric Transformations II*, New Mathematical Library, Vol. 21, Random House, 1968.

74. Yaglom, I. M., *Geometric Transformations III*, New Mathematical Library, Vol. 24, Random House, 1973.

75. Zeitz, P., *The Art and Craft of Problem Solving*, John Wiley & Sons, 1999.

Afterword

This book is the product of many years of work and is based on the authors' extensive experience in mathematics and mathematical education. Both authors have extensive experience in the development and composition of original mathematics problems and in the applications of advanced methodologies in mathematical science teaching and learning.

This book is aimed at three major types of audiences:

(a) Students ranging from high school juniors to college seniors. This book will prove useful to those wanting to tie up many loose ends in their study of combinatorics and to develop mathematically, in general. Students with interest in mathematics competitions should have this book in their personal libraries.

(b) Numerous teachers who are implementing problem solving across the nation. This book is a perfect match for teachers wanting to teach advanced problem-solving classes and to organize mathematical clubs and circles.

(c) Amateur mathematicians longing for new mathematical gems and brain teasers. This book presents sophisticated applications of genuine mathematical ideas in real-life examples. It will help them to recall the experience of reading the wonderful stories by Martin Gardner in his monthly column in *Scientific American*.

By studying this book, readers will be well-equipped to further their knowledge in more abstract combinatorics and its related fields in mathematical and computer science. This book serves as a solid stepping stone for advanced mathematical reading, such as *Combinatorial*

Theory by M. Aigner, *Concrete Mathematics — A Foundation for Computer Science* by R. L. Graham; D. E. Knuth; and O. Patashnik, and *Enumerative Combinatorics* by R. P. Stanley.